GREAT MOMENTS
IN
MATHEMATICS
(BEFORE 1650)

By
HOWARD EVES

THE
DOLCIANI MATHEMATICAL EXPOSITIONS

Published by
THE MATHEMATICAL ASSOCIATION OF AMERICA

The Dolciani Mathematical Expositions

NUMBER FIVE

GREAT MOMENTS IN MATHEMATICS
(BEFORE 1650)

By
HOWARD EVES
University of Maine

Published and distributed by
THE MATHEMATICAL ASSOCIATION OF AMERICA

© 1980 by
The Mathematical Association of America (Incorporated)
Library of Congress Catalog Card Number 80-81046

Complete Set ISBN 0-88385-300-0
Vol. 5 ISBN 0-88385-305-1

Printed in the United States of America

Current printing (last digit):
10 9 8 7 6 5 4 3 2 1

The DOLCIANI MATHEMATICAL EXPOSITIONS series of the Mathematical Association of America was established through a generous gift to the Association from Mary P. Dolciani, Professor of Mathematics at Hunter College of the City University of New York. In making the gift, Professor Dolciani, herself an exceptionally talented and successful expositor of mathematics, had the purpose of furthering the ideal of excellence in mathematical exposition.

The Association, for its part, was delighted to accept the gracious gesture initiating the revolving fund for this series from one who has served the Association with distinction, both as a member of the Committee on Publications and as a member of the Board of Governors. It was with genuine pleasure that the Board chose to name the series in her honor.

The books in the series are selected for their lucid expository style and stimulating mathematical content. Typically, they contain an ample supply of exercises, many with accompanying solutions. They are intended to be sufficiently elementary for the undergraduate and even the mathematically inclined high-school student to understand and enjoy, but also to be interesting and sometimes challenging to the more advanced mathematician.

———

The following DOLCIANI MATHEMATICAL EXPOSITIONS have been published.

Volume 1: MATHEMATICAL GEMS, by Ross Honsberger

Volume 2: MATHEMATICAL GEMS II, by Ross Honsberger

Volume 3: MATHEMATICAL MORSELS, by Ross Honsberger

Volume 4: MATHEMATICAL PLUMS, edited by Ross Honsberger

Volume 5: GREAT MOMENTS IN MATHEMATICS (BEFORE 1650), by Howard Eves

A Grateful Acknowledgment

I wish to express my sincerest thanks to Professor Ross Honsberger and Professors G. L. Alexanderson, J. Malkevitch, and K. R. Rebman, Chairman and Members of the Dolciani Subcommittee, for their careful and constructive reading of the original manuscript. Their criticisms and suggestions were invaluable.

HOWARD EVES

Dedicated to Carroll V. Newsom,
in friendship and gratitude

PREFACE

Many who were growing up in America during the period from 1928 to 1942 will recall the extreme pleasure they experienced when they tuned in their radios to the NBC "Music Appreciation Hour." During those fourteen years an estimated five million school children and a large unknown number of just plain devotees listened each week to the beautifully modulated and hypnotic voice of Walter Damrosch, eminent musicologist and master popularizer of good music. Week after week the listeners auricularly drank in the "great moments in music," and, ever after, their lives were enriched by knowing something of the noble accomplishments of the world's famous composers.

Some years later, while serving in the Mathematics Department of Oregon State College, it occurred to me that what Walter Damrosch so magnificently did for music could perhaps also be done for mathematics. Why not, I thought, develop a set of lectures devoted to the enthralling GREAT MOMENTS IN MATHEMATICS? I reacted to the idea with enthusiasm. I would aim the lectures, I decided, at a more specific audience than that chosen by Walter Damrosch—I had in mind, of course, a college and college-community audience. My major hope was to reach, without any great mathematical demands, anyone interested in learning something about the outstanding achievements in mathematics over the ages. The whole thing was to be an intellectual adventure, with no truly prohibiting or frightening prerequisites. And yet at the same time I also wanted to give something that would challenge a good mathematics student and perhaps be of use to teachers of mathematics.

The somewhat conflicting aims of the lecture series were finally met as follows. A lecture sequence of some sixty chronologically ordered GREAT MOMENTS IN MATHEMATICS was designed, along with ample problem material, sometimes in the form of "junior" research, bearing on the subject matter of the lectures. The lectures, each fifty minutes long, and enough of them to carry through two semesters, were to be offered at two meetings each week, and the associated problem material was to be discussed at a third weekly meeting. The two-meeting sequence was to constitute an appreciation course, open to auditors and to seekers of a total of four elective college credits; for the first semester (which is covered in the present volume) an acquaintance with high school mathematics was the only prerequisite. The three-meeting sequence was to constitute a mathematics course, open to qualified students and teachers seeking a total of six college credits in mathematics; somewhat stiffer mathematical demands—say, mathematics through beginning calculus—were made of those registering for the extended sequence.

Upon being invited by the Publication Committee for the Dolciani Mathematical Expositions to write up the GREAT MOMENTS IN MATHEMATICS lecture series, I decided, because of space problems, to attempt a curtailed version. Only forty of the lectures were selected, the first twenty from the period before 1650, and the remaining twenty from the period after 1650. Herewith are the first twenty.

Each selected lecture has been mercilessly pared down, inasmuch as a transcript of a complete fifty-minute lecture would run into far too many pages. Thus, almost all of the humor and anecdotal material so fitting in an oral presentation, as well as many of the cultural ramifications and side trails, and, of course, all the visual props in the form of models, displays, maps, portraits, and overhead-projector material, are omitted.

Consider, for example, LECTURE 9 of the curtailed series, devoted to Archimedes and his method of equilibrium. Recently, at an oral presentation of this lecture, I had at the lecture desk a reproduction of an ancient Greek sand tray, a specimen of a palimpsest, an attractively boxed *loculus Archimedius*, a small demonstration model of an Archimedean screw, a large calibrated circular cylinder with a heavy removable inscribed sphere, working models of the three classes of levers for comparing their mechanical advantages, and a compound pulley attached to a heavy weight, which was almost effortlessly moved every now and then during the lecture. I showed, on an overhead projector, transparencies of the three questionable medallion portraits of Archimedes, a picture of the

interesting mosaic portraying Archimedes' last moments now re-
siding in the Municipal Art Institute at Frankfurt am Main, a
portrait of Heiberg, a picture of a sculptured bust of Marcellus,
and a map of ancient Syracuse. To lighten the oral presentation I
introduced bits of humor—bits which might appear somewhat
ridiculous if reproduced here in print. But, as I learned years ago
at Harvard from my mentor Julian Lowell Coolidge, a touch of
clownery can have a place in an oral presentation. The mathematical demonstration in the lecture, which appears so terse and stark
in the written version, was carefully, slowly, and meticulously performed at a blackboard, so that the audience could almost *see* the
elements of volume being slid to their appropriate positions along
the balance bar. And, along with all the shortcuts and omissions,
the lyrical and poetical flights of an oral presentation are also miss-
ing in the written version. What has been said of LECTURE 9 can
also be said—sometimes, it is true, not so fully—of each of the
other lectures of the series.

So, here, with sincere apologies, are cruelly condensed versions
of some of the lectures on the GREAT MOMENTS IN MATHEMATICS.
It could be that the only proper way to preserve the lectures would
be on videotape, or, better, on educational TV, delivered by a
gifted lecturer and with all the props and marvels possible with
such a presentation.

A few closing words are perhaps in order. The selection of the
GREAT MOMENTS is, of course, my own, and could well differ from
a selection made by someone else. Some of the GREAT MOMENTS
can be precisely pinpointed in the time strip—others only vaguely.
It must also be remembered that a *moment* in history is sometimes
an inspired flash and sometimes an evolution extending over a long
period of time. Much of the subject matter and many of the problems of the lectures subsequently found a place in my *Introduction
to the History of Mathematics* and in *An Introduction to the Foundations and Fundamental Concepts of Mathematics*, which I wrote
with Carroll V. Newsom, and the anecdotes and stories, considerably augmented, now appear in my four *Mathematical Circles*
books. Finally, in a few spots, for the sake of brevity and to avoid
complexities beyond the scope of the lectures, certain minor simpli-

fications have been introduced that are hoped to be essentially unimportant so far as the purpose and the honesty of the lectures are concerned.

HOWARD EVES

Fox Hollow, Lubec, Maine
Winter 1977-78

CONTENTS

xiii

SCRATCHES AND GRUNTS

In the Homeric legends it is narrated that when Ulysses left the land of the Cyclops, after blinding the one-eyed giant Polyphemus, that unfortunate old giant would sit each morning by the entrance to his cave and from a heap of pebbles would pick up one pebble for each ewe that he let pass out of the cave. Then, in the evening, when the ewes returned, he would drop one pebble for each ewe that he admitted to the cave. In this way, by exhausting the supply of pebbles he had picked up in the morning, he was assured that all his flock had returned in the evening.

The story of Polyphemus is one of the earliest literary references to the notion of a one-to-one correspondence as the basis of counting. Many illustrations of the principle involved can be given. Thus, on a somewhat gruesome note, certain American Indians kept count of the number of enemies slain by collecting the scalp of each vanquished foe, and certain primitive African hunters, in proving their manhood, still keep count of the number of wild boars killed by collecting the tusks of each animal. The young unmarried girls of the tribe of Masai herdsman who live on the slopes of Mt. Kilimanjaro used to wear a number of brass rings about their necks equal to their ages. The English idiom "to chalk one up" arose from the custom of early bartenders keeping count of a customer's drinks by making chalk marks on a slate, and the Spanish idiom "echai chinas" ("to toss a pebble") arose from the similar custom of early Spanish bartenders keeping count by tossing pebbles in the customer's hood. It is also apparent in the body counting of some primitive peoples, wherein certain parts of the body are used to indicate various numbers. It is again apparent in the once widespread use of tally sticks, in which accounts were recorded by appropriate notches cut in pieces of wood; the tally sticks employed by the British Exchequer remained legal registers until as late as 1826. The ancient Peruvians maintained population and other counts on a *quipu*—a device consisting of a cord with attached knotted strings of various colors. And, of course, children today keep count of the days till Christmas or a vacation from school by checking the days off on a calendar. Almost anyone will, at one time or another, keep a small tally by ticking off on his fingers.

The oldest extant artifact of mathematical significance is a bone tool-handle, bearing notches arranged in definite numerical patterns, with a piece of quartz fitted into a narrow cavity at the head of the handle. Known as the *Ishango bone*, it was found in 1962 by Jean de Heinzelin at the fishing site of Ishango, on the shore of Lake Edward in the Democratic Republic of the Congo, and dates back to the period between 9000 and 6500 B.C. The meaning of the tally notches can only be conjectured, and there is a difference of opinion among the examining experts.

Quite likely the earliest GREAT MOMENT IN MATHEMATICS occurred when, many thousands of years ago, primitive man began to keep count of certain collections by making scratches in the dirt or on a stone. Society had evolved to the point where simple counting became imperative. A tribe, a clan, or a family had to apportion food among its members, or had to keep track of the size of a flock or herd. The process was a simple tally method employing the principle of one-to-one correspondence and was probably the beginning of the science of writing.

It seems fair to surmise that in keeping a count of a small collection, one finger was either raised or turned down per member of the collection. Tally counts for larger collections could, as indicated by the examples above, be made by assembling pebbles or sticks, by making scratches in the dirt or on a stone, by cutting notches on a bone or in a piece of wood, or by tying knots in a string. Perhaps later an assortment of grunts was developed as a vocal tally against the number of objects in small collections. Still later, an assortment of written symbols (*numerals*) was evolved to represent these numbers.

Although this development of early counting is largely conjectural, it is supported by reports of anthropologists in their studies of present-day primitive peoples and by certain artifacts unearthed in various parts of the world. It is the way small children of today begin to keep count.

In the earlier stages of the vocal period of counting, different grunts (words) were used, for example, for *two* sheep and *two* men. One merely has to recall that in English we still use *team* of horses, *span* of mules, *yoke* of oxen, *brace* of partridge, *pair* of shoes. The ultimate abstraction of the common property of *two*, represented by some sound considered independently of any concrete association, probably was a long time in arriving. Our present number words in all likelihood originally referred to sets of certain concrete objects, but these connections, except for that perhaps relating *five* and *hand*, are now lost to us.

The relation of certain number words to a concrete tally association still lingers in some primitive societies of today. Thus, because of a peculiar system of counting among a Papuan tribe in southeast New Guinea, it was found necessary to translate the Bible passage (John 5:5): "And a certain man was there, which had an infirmity thirty and eight years" into "A man lay ill one man (20), both hands (10), 5 and 3 years." Again, since primitive peoples count on their fingers, sometimes the names of the fingers are actually used by the people as number words. Thus the South American Kamayura tribe use the word "peak-finger" (middle finger) as their word for "three," and "three days" comes out as "peak-finger days." Again, the Dene-Dinje Indians of South America, who count by successively folding down the fingers of their hands, count by the following literal equivalents:

"one"—"the end is bent" (the little finger is folded)
"two"—"it is bent once more" (the ring finger is also folded)
"three"—"the middle is bent" (the middle finger is also folded)
"four"—"only one remains" (only the thumb is still extended)
"five"—"my hand is ended" or "my hand is dead" (all fingers and thumb are folded)
"ten"—"my hands are dead"
"four days"—"only-one-remains days"

Interesting is the word *kononto*, for "nine," of the Mandingo tribe of West Africa; the word literally means "to the one in the belly"—

a reference to the nine months of pregnancy. The concrete stage in counting is also evident in the Malay and Aztec tongues, where the numbers "one," "two," "three" are, literally, "one stone," "two stones," "three stones." Similarly, among the Niuès of the Southern Pacific, the first three number words are, literally, "one fruit," "two fruits," "three fruits," and among the Javanese they are, literally, "one grain," "two grains," "three grains."

There are instances where a silent language, in the form of appropriate gestures, may be employed in the one-to-one correspondence used for counting. Thus there is a Papuan body counting wherein to indicate small numbers one touches the appropriate part of the body according to the following scheme:

1	right little finger	12	nose
2	right ring finger	13	mouth
3	right middle finger	14	left ear
4	right index finger	15	left shoulder
5	right thumb	16	left elbow
6	right wrist	17	left wrist
7	right elbow	18	left thumb
8	right shoulder	19	left index finger
9	right ear	20	left middle finger
10	right eye	21	left ring finger
11	left eye	22	left little finger

One notes the mirrorlike repetition in reverse, interrupted by "nose" and "mouth" for 12 and 13.

It is common among primitive people, and even among sophisticated people, to accompany verbal counting with gestures. For example, in some tribes the word "ten" is frequently accompanied by clapping one hand against the palm of the other, and the word "six" is sometimes accompanied by passing one hand rapidly over the other. Karl Menninger says that certain African tribes can be identified and ethnically classified by observing whether they begin to count on the left hand or the right hand, whether they unfold the fingers or bend them in, or whether they turn the palm toward the body or away from the body.

The Englishman R. Mason has related a charming anecdote about World War II. A Japanese girl was in India, which at the time was at war with Japan. To avoid a possibly embarrassing situation, her friend introduced her as Chinese to an English resident of India. The Englishman was skeptical and asked the girl to count to five on her fingers, which, after some hesitation, she did. Then:

> Mr. Headley burst out delightedly: "There you are! Did you see that? Did you see how she did it? Began with her hand open and bent her fingers in one by one. Did you ever see a Chinese do such a thing? Never! The Chinese count like the English. Begin with the fist closed. She's Japanese!" he cried triumphantly.

The notion of one-to-one correspondence has long been realized as the basis for counting finite collections. In an extraordinary series of articles, beginning in 1874 and published for the most part in the mathematics journals *Mathematische Annalen* and *Journal für Mathematik*, the German mathematician Georg Cantor applied the same basic notion to the counting of infinite collections, and thereby created the remarkable theory of *transfinite numbers*. But this is another, and of course much more recent, GREAT MOMENT IN MATHEMATICS; it will be properly considered in its own place in a later lecture.

Exercises

1.1. Explain the Papuan translation of the Bible passage John 5:5 cited in the lecture text.

1.2. Explain how "peak-finger" became the word for "three" among the Kamayura tribe of South America.

1.3. The Zulus of South Africa use the following equivalents:

> "six"—"taking the thumb"
> "seven"—"he pointed"

Can you furnish an explanation for this?

1.4. The Malinké of West Sudan use the word *dibi* for "forty." The word literally means "a mattress." Can you give an explanation for this?

1.5. In British New Guinea, the number "ninety-nine" comes out as "four men die, two hands come to an end, one foot ends, and four." Explain this.

1.6. Two sets are said to be *equivalent* if and only if they can be placed in one-to-one correspondence. Show that

(a) the set of all letters of the alphabet is equivalent to the set of the first 26 positive integers;

(b) the set of all positive integers is equivalent to the set of all even positive integers;

(c) equivalence of sets is reflexive, symmetric, and transitive.

1.7. Two sets that are equivalent are said to have the *same cardinal number*. Let A be a set of cardinal number α and B a set of cardinal number β, where A and B have no element in common. Then, by $\alpha + \beta$, called the *sum* of α and β, we mean the cardinal number of the set $A \cup B$. This binary operation on cardinal numbers is called *addition*. Prove that addition of cardinal numbers is commutative and associative.

1.8. The set C whose elements are all ordered pairs (a, b), where a is an element of set A and b is an element of set B, is called the *Cartesian product* of A and B, and is denoted by $A \times B$. If A has cardinal number α and B has cardinal number β, then, by $\alpha\beta$, called the product of α and β, we mean the cardinal number of the set $C = A \times B$. This binary operation on cardinal numbers is called *multiplication*. Prove that multiplication of cardinal numbers is commutative, associative, and distributive over addition of cardinal numbers.

1.9. Show that a set A consisting of five elements contains 2^5 subsets (including itself and the null set). Generalize to the case of any finite set A.

1.10. Let A be a set with seven elements and B a set with five elements. What can be said about the number of elements in the sets $A \cap B$ and $A \cup B$? Generalize to the case of any two finite sets A and B.

Further Reading

MENNINGER, KARL, *Number Words and Number Symbols, a Cultural History of Numbers.* Cambridge, Mass.: The M.I.T. Press, 1969.

ZASLAVSKY, CLAUDIA, *Africa Counts, Numbers and Patterns in African Culture.* Boston: Prindle, Weber & Schmidt, 1973.

THE GREATEST EGYPTIAN PYRAMID

The first geometrical considerations of man must be very ancient and must have subconsciously originated in simple observations stemming from human ability to recognize physical form and to compare shapes and sizes. Certainly one of the earliest geometrical notions to thus impinge itself on even the least reflective mind would be that of distance, in particular the concept that the straight line is the shortest path connecting two points; for most animals seem instinctively to realize this. Another early notion that would gradually emerge from the subconscious to the conscious mind would be that of simple rectilinear forms, such as the triangle and the quadrilateral. Indeed, it seems almost instinctive in laying out boundaries first to locate the corners and then to connect the successive corners by straight-line walls or fences. In building walls the notions of vertical, parallel, and perpendicular would gradually emerge. Many special curves, standing out among the generally haphazard shapes of nature, would impress themselves on man's subconscious mind. Thus the discs of the sun and full moon are circular, as is an arc of a rainbow and the cross-section of a log. The parabolic trajectory of a hurled stone, the catenary curve of a hanging vine, the spiral curve of a coiled rope, and the helical curve of certain tendrils would similarly be noticed by even the least observant mind. Certain spiders spin webs that closely approximate regular polygons. The swelling set of concentric circles caused by a stone cast into a pond and the attractive flutings on many shells suggest families of associated curves. Many fruits and pits are spherical; tree trunks are circular cylinders; conical shapes appear here and there in nature. Surfaces and solids of revolution, observed in nature, as among melons, or from work on a potter's

8

wheel, would subconsciously strike an inquisitive mind. Man, animals, and many leaves possess a bilateral symmetry. The notion of volume would be encountered every time a container was filled at the spring or river bank. The conception of space and of points in space is involved whenever one looks at the stars in the sky at nighttime. The list is easily extended.

This first nebulous acquaintance with many geometrical concepts may be called *subconscious geometry*. It was employed by early peoples, as it is by children today, in their primitive art work.

The second stage in geometry arose when human intelligence was able to extract from a set of concrete geometrical relationships a general abstract relationship containing the former as particular cases. One thus arrives at a geometrical law or rule. For example, in measuring the areas of various rectangles drawn upon quadrille-ruled paper by counting the number of little squares of the paper found inside the rectangles, a young grade-school pupil would soon induce that the area of any rectangle is probably given by the product of its two dimensions. Again, in measuring, by a tape measure, the circumferences of a number of wooden circular discs, the young pupil would induce that the circumference of any circle is somewhat more than three times the diameter of the circle.

As a more sophisticated example, consider a horizontal wooden circular disc with an upright nail driven part way into its center, and a wooden hemisphere of the same radius as the disc with a nail driven part way into its pole. Now coil a thick cord on the disc, in spiral fashion from the nail, until the disc is covered, noting the length of cord required to do this. Next coil the same kind of cord in spiral fashion about the nail in the hemisphere until the hemisphere is covered, again noting the length of cord required. In comparing the lengths of the cords used for the disc and for the hemisphere, it will be found that the latter is always (very closely) twice the former. From this one could induce that the area of the hemisphere is twice that of the disc, or that the area of a sphere is equal to four times the area of one of its great circles—a fact that was first rigorously established by Archimedes in the third century B.C. With such experiments, geometry became a laboratory study.

The laboratory stage in geometry is known as *scientific* (or *experimental*, or *empirical*, or *inductive*) *geometry*. As far back as history allows us to grope into the past, we find already present a sizable body of scientific geometry. This type of geometry seems to have arisen in certain advanced pockets of the ancient Orient (the world east of Greece) in the fifth to the third millennium B.C., to assist in engineering, agricultural, and business pursuits, and in religious ritual.

It is interesting that all recorded geometry prior to 600 B.C. is essentially scientific geometry. Geometry developed into a large bundle of rules of thumb, some correct and some only approximately correct. In a course in the history of mathematics, considerable attention is given to examining the laboratory nature of the geometry of the ancient Babylonians, Egyptians, Hindus, and Chinese. To illustrate, consider an early Chinese formula for the area of a segment of a circle. The formula is found in the *Arithmetic in Nine Sections*, dating from the second century B.C. but, because of the burning of the books in 213 B.C., believed to be a restoration of a much earlier work. In Figure 1, let c represent the chord and s the sagitta* of the circular segment. If from the midpoint of the arc of the segment one draws secants cutting the extensions of c so that the extended parts are each equal to half of s, our eyes tell us that the circular segment is approximately equal in area to the isosceles triangle formed by the line of c and the two secant lines. Assuming the areas are actually equal, we find the old Chinese formula $A = s(c + s)/2$ for the area of the circular segment. Applying this to a semicircular segment, it is easily shown that, in this case, the formula is equivalent to taking $\pi = 3$, an approximation of π frequently found in ancient mathematics.

In the Rhind papyrus, an Egyptian work on mathematics dating back at least to 1650 B.C., we find the area of a circle taken as equal to that of a square having eight-ninths of the circle's diameter as a side. It can be shown that this empirical formula is equivalent to taking $\pi = (4/3)^4 = 3.1604 \ldots$.

Although most of the baked clay mathematical tablets lifted in Mesopotamia show that the ancient Babylonians took $\pi = 3$, a

---

*The distance from the midpoint of the chord of the segment to the midpoint of the arc of the segment.

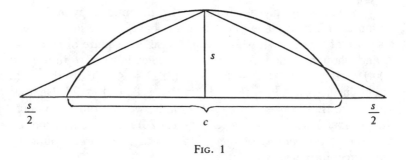

recently discovered tablet dating from 1900 to 1600 B.C. and unearthed in 1936 at Susa, about 200 miles from Babylon, gives the
better estimate of $3\frac{1}{8} = 3.125$.

Many other examples of the scientific nature of very early geometry are known. One is impressed by the amount of geometry that
can be discovered by purely laboratory methods.

If, from the accumulation of examples of scientific geometry
that have come down to us from antiquity, one were to pick an
outstanding instance that might serve as a GREAT MOMENT IN
MATHEMATICS, one could scarcely do better than to settle on Problem 14 of the Moscow papyrus. The Moscow papyrus, dating back
to approximately 1850 B.C., is a mathematical text containing 25
problems which were already old when the manuscript was com-
piled. The papyrus was purchased in Egypt in 1893 and now re-
sides in a museum in Moscow. In Problem 14 of the papyrus we
find the following numerical example:

"You are given a truncated pyramid of 6 for the vertical height
by 4 on the base by 2 on the top. You are to square this 4, result
16. You are to double 4, result 8. You are to square 2, result 4.
You are to add the 16, the 8, and the 4, result 28. You are to take
one-third of 6, result 2. You are to take 28 twice, result 56. See, it
is 56. You will find it right."

Now what are we to make of this? First of all we are to realize
that, following the custom in illustrative problems in antiquity, a
general procedure is being described and the specific numbers employed are only incidental. Since all the extant Egyptian pyramids
of ancient times are regular square pyramids, we assume that in

the problem we are given a frustum of a pyramid (a pyramid with its top cut off by a plane parallel to the base) whose lower base is a square of side $a = 4$, whose upper base is a square of side $b = 2$, and whose altitude is $h = 6$. We are told, in turn, to find $a^2 = (4)(4) = 16$, $ab = (4)(2) = 8$, $b^2 = (2)(2) = 4$. We are then told to obtain the sum $a^2 + ab + b^2 = 16 + 8 + 4 = 28$. Next we are told to find $\frac{1}{3}h = \frac{1}{3}(6) = 2$. Finally, we are told to compute the product $\frac{1}{3}h(a^2 + ab + b^2) = (2)(28) = 56$. This product, however, is the *volume* of the given truncated pyramid found by the correct formula

$$V = \frac{1}{3}h(B_1 + \sqrt{B_1 B_2} + B_2)$$

for the volume of any frustum of a pyramid with lower base of area B_1, upper base of area B_2, and altitude h.

Let us pause a moment to consider, assuming our interpretation of Problem 14 is correct, the remarkableness of the above. The ancient Babylonians knew that the area of a trapezoid (which can be regarded as a truncated triangle) is given by the product of its altitude and half the sum of its two bases. Analagous to this, the ancient Babylonians took the volume of a frustum of a pyramid as the product of the altitude of the frustum and half the sum of the areas of its two bases, or, in the notation introduced above,

$$V = \frac{1}{2}h(B_1 + B_2).$$

Now, though it is natural to conjecture that this formula yields the volume of the frustum, the formula is incorrect. To find the volume of the frustum we expect, of course, to multiply the altitude h by some sort of mean or average of the areas B_1 and B_2. But the *arithmetic* mean of B_1 and B_2, namely, $\frac{1}{2}(B_1 + B_2)$, is not correct. What one needs here (and which is not at all obvious) is the *heronian* mean of B_1 and B_2, namely,

$$\frac{1}{3}(B_1 + \sqrt{B_1 B_2} + B_2).$$

The ancient Egyptian author of Problem 14 of the Moscow papyrus, somehow or other and unlike the ancient Babylonians, made the correct conjecture. Surely this induction is a truly remarkable piece of empirical work in geometry. So remarkable did it seem to Eric Temple Bell that he named Problem 14 of the Moscow papyrus "the greatest Egyptian pyramid"; to Bell, the induction involved in the problem is far more remarkable than the actual physical construction of any of the massive stone pyramids of Egyptian antiquity still standing today. It was a GREAT MOMENT IN MATHE-MATICS.

Exercises

2.1. (a) Follow through the empirical procedure, described in the lecture text, leading to the old Chinese formula for the area of a segment of a circle.

(b) Show that applying the formula to a semicircular segment is equivalent to taking $\pi = 3$.

(c) Derive a correct formula for the area of a circular segment in terms of the chord c and the sagitta s of the segment.

2.2. (a) Show that the ancient Egyptian method of finding the area of a circle is equivalent to taking $\pi = (4/3)^4 = 3.1604 \ldots$.

(b) Form an octagon from a square of side 9 units by trisecting the sides of the square and then cutting off the four triangular corners. The area of the octagon looks, by eye, to differ very little from the area of the circle inscribed in the square. Show that the area of the octagon is 63 square units, whence the area of the circle cannot be far from that of a square of 8 units on a side. There is evidence, in the form of a crudely drawn figure accompanying Problem 48 of the Rhind papyrus, that the Egyptian formula for the area of a circle may have been arrived at in this way.

2.3. On an old Babylonian baked clay tablet lifted at Susa in 1936, the ratio of the perimeter of a regular hexagon to the circumference of the circumscribed circle is given as $57/60 + 36/3600$. Show that this leads to $3\frac{1}{8}$ as an approximation of π.

2.4. The idea of averaging is common in empirical work. Thus we find, in the Rhind papyrus, the area of a quadrilateral having successive sides a, b, c, d given by

$$K = \left(\frac{a + c}{2}\right)\left(\frac{b + d}{2}\right).$$

(a) Show that, actually, the formula above gives too large a result for all nonrectangular quadrilaterals.

(b) If the Egyptian formula above is assumed correct, show that the area of a triangle would be given by half the sum of two sides multiplied by half the third side. We find this incorrect formula for the area of a triangle in an extant deed from Edfu dating some 1500 years after the Rhind papyrus.

2.5. Interpret the following, found on a Babylonian tablet believed to date from about 2600 B.C.:

"60 is the circumference, 2 is the sagitta, find the chord.

"Thou, double 2 and get 4, dost thou not see? Take 4 from 20, thou gettest 16. Square 20, thou gettest 400. Square 16, thou gettest 256. Take 256 from 400, thou gettest 144. Find the square root of 144. 12, the square root, is the chord. Such is the procedure."

2.6. The *Śulvasūtras*, ancient Hindu religious writings dating from about 500 B.C., are of interest in the history of mathematics because they embody certain geometrical rules for the construction of altars and show an acquaintance with the Pythagorean theorem. Among the rules furnished there appear empirical solutions to the circle-squaring problem which are equivalent to taking $d = (2 + \sqrt{2})s/3$ and $s = 13d/15$, where d is the diameter of the circle and s is the side of the equivalent square. These formulas are equivalent to taking what values for π?

2.7. If m and n are two positive numbers, we define the *arithmetic mean*, the *heronian mean*, and the *geometric mean* of m and n to be $A = (m + n)/2$, $H = (m + \sqrt{mn} + n)/3$, $G = \sqrt{mn}$. Show that $A \geq H \geq G$, the equality signs holding if and only if $m = n$.

2.8. Assuming the familiar formula for the volume of any pyramid (volume equals one-third the product of base and altitude), show that the volume of any frustum T of a pyramid is given by the product of the height of T and the heronian mean of the two bases of T.

2.9. Let a, b, and h denote the lengths of an edge of the lower base, an edge of the upper base, and the altitude of a frustum T of a regular square pyramid. Dissect T into: (1) a rectangular parallelepiped P of upper base b^2 and altitude h, (2) four right triangular prisms A, B, C, and D each of volume $b(a - b)h/4$, (3) four square pyramids E, F, G, and H each of volume $(a - b)^2h/12$. Now obtain the formula

$$V = h(a^2 + ab + b^2)/3$$

for the volume of T.

2.10. Consider the dissected frustum T of Exercise 2.9. Horizontally slice P into three equal parts each of altitude $h/3$ and designate one of these slices by U. Combine A, B, C, D into a rectangular parallelepiped Q of base $b(a - b)$ and altitude h, and horizontally slice Q into three equal parts of altitude $h/3$. Replace E, F, G, H by a rectangular parallelepiped R of base $(a - b)^2$ and altitude $h/3$. Combine one slice of P with one slice of Q to form a rectangular parallelepiped V of base ab and altitude $h/3$. Combine one slice of P, two slices of Q, and R to form a rectangular parallelepiped W of base a^2 and altitude $h/3$. The volume of T is then equal to the sum of the volumes of the three rectangular parallelepipeds U, V, W. Using this fact find the formula of Exercise 2.9 for the volume of T. It has been suggested that the procedure in Problem 14 of the Moscow papyrus may have been obtained in this fashion.

Further Reading

GILLINGS, R. J., *Mathematics in the Time of the Pharaohs*. Cambridge, Mass.: The M.I.T. Press, 1972.

NEUGEBAUER, OTTO, *The Exact Sciences in Antiquity*, 2nd ed. New York: Harper & Row, 1962.

FROM THE LABORATORY INTO THE STUDY

It was about 600 B.C. that geometry entered a third stage of development. Historians of mathematics are unanimous in accrediting this further advancement to the Greeks of the period, and the earliest pioneering efforts to Thales of Miletus, one of the "seven wise men" of antiquity. Thales, it seems, spent the early part of his life as a merchant, becoming wealthy enough to devote much of his later life to study and some travel. He visited Egypt and brought back with him to Miletus knowledge of Egyptian accomplishments in geometry. His many-sided genius won him a reputation as a statesman, counselor, engineer, business man, philosopher, mathematician, and astronomer. He is the first individual known by name in the history of mathematics, and the first individual with whom deductive geometrical discoveries are associated. He is credited with the following elementary results:

1. A circle is bisected by any diameter.
2. The base angles of an isosceles triangle are equal.
3. Vertical angles formed by two intersecting lines are equal.
4. Two triangles are congruent if they have two angles and one side in each respectively equal.
5. An angle inscribed in a semicircle is a right angle.

Now all five of the results above were undoubtedly known long before Thales' time, and all five are easily arrived at in a laboratory. So the value of these results is not to be measured by their content, but rather by the belief that Thales supported each of them by some logical reasoning instead of by intuition and experiment. Take, for example, the third result, which, in the laboratory, would easily be verified by cutting out a pair of vertical

16

angles with scissors and applying one of the angles to the other. Thales, however, probably reasoned out the result much as we do today in a beginning geometry class. In Figure 2, we want to show that angle x = angle y. Now angle x is the supplement of angle z; also, angle y is the supplement of angle z. Therefore, since things equal to the same thing are equal to one another, it follows that angle x = angle y. The desired result has been obtained by a small chain of deductive reasoning, stemming from a more fundamental result. This type of geometry is known as *demonstrative* (or *deductive*, or *systematic*) *geometry*, and was considerably developed by the Greeks from 600 B.C. on. These early Greeks removed the establishment of geometrical, and similarly all mathematical, results from the laboratory into the study. This conscious and deliberate effort was certainly a GREAT MOMENT IN MATHEMATICS, and, if tradition is correct, Thales of Miletus was the original motivator.

Just why, of all the peoples of the time, the Greeks decided that geometrical facts must be assured by logical demonstration rather than by laboratory experimentation is sometimes referred to as the *Greek mystery*. Scholars have tried to furnish explanations of the Greek mystery, and though no one explanation by itself seems wholly satisfying, it may be that all of them together are acceptable. The most commonly given explanation finds the reason in the peculiar mental bias of the Greeks of classical times toward philosophical inquiries. In philosophy one is concerned with

Euclid , 115

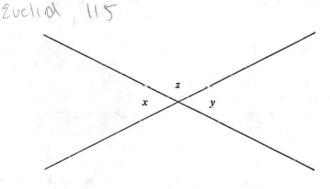

FIG. 2

inevitable conclusions that follow from assumed premises, and the empirical method affords only a measure of probability in favor of a given result. It is deductive reasoning that philosophers find to be their indispensable tool, and so the Greeks naturally gave preference to this method when they began to consider geometry.

Another explanation of the Greek mystery lies in the Hellenic love of beauty, as is manifest in their art, their writing, their sculpture, and their architecture. Now appreciation of beauty is an intellectual as well as an emotional experience, and from this point of view the orderliness, the consistency, the completeness, and the conviction found in deductive argument are very satisfying.

A still further explanation of the Greek mystery has been found in the slave-based nature of Greek society in classical times. The privileged class was supported by a large slave class that ran the businesses, managed the industries, took care of the households, and did both the technological and the unskilled work of the time. This slave basis naturally fostered a separation of theory from practice and led members of the privileged class to a preference for deduction and abstraction and a disdain for experimentation and practical application.

Finally, the explanation may lie essentially in the sweeping economic and political changes that occurred at the time. The Iron Age had been ushered in, the alphabet had been invented, coins were introduced, and geographic discoveries were made. The world was ready for a new type of civilization, and this new civilization made its appearance among the more forward-looking and imaginative people in the trading towns that sprang up along the coast of Asia Minor and later on the mainland of Greece, in Sicily, and on the Italian shore. These trading towns were largely Greek settlements. Under a developing atmosphere of rationalism, men began to ask *why* as well as *how*. Now empirical processes are quite adequate for the question how, but they do not suffice to answer inquiries of why, and attempts at demonstrative methods were bound to assert themselves, with the result that the deductive feature, which modern scholars regard as a fundamental characteristic of mathematics, came into prominence.

But whatever may be the true explanation of the Greek mystery, it must be conceded that the Greeks of classical times converted geometry into something vastly different from the collection of empirical rules handed down by their predecessors. Moreover, the fact that the first deductive thinking was done in the field of geometry instead of algebra inaugurated a tradition in mathematics that was maintained until quite recent times.

It must not be thought that the Greeks shunned all preliminary empirical and experimental methods in mathematics, for it is probably quite true that few, if any, significant mathematical facts have ever been found without some preliminary empirical work of one form or another. Before a mathematical statement can be proved or disproved by deduction, it must first be conjectured, and a conjecture is nothing but a guess made more or less plausible by intuition, observation, analogy, experimentation, or some other form of empirical procedure. Deduction is a convincing formal mode of exposition, but it is hardly a means of discovery. It is a set of complicated machinery that needs material to work upon, and the material is usually furnished by empirical considerations. Even the steps of a deductive proof or disproof are not dictated to us by the deductive apparatus itself but must be arrived at by trial and error, experience, and shrewd guessing. Indeed, skill in the art of good guessing is one of the prime ingredients in the makeup of a worthy mathematician. What is important here is that the Greeks insisted that a conjectured or laboratory-obtained mathematical statement must be followed up with a rigorous proof or disproof by deduction and that no amount of verification by experiment is sufficient to *establish* the statement.

To succeed in geometry, either as a creator or simply as a problem-solver, one must be willing to experiment, to draw and test innumerable figures, to try this and to try that. Galileo (1564–1642), in 1599, attempted to ascertain the area under one arch of the cycloid curve* by balancing a cycloidal template against circular templates of the size of the generating circle. Because of a slight flaw in his platform balance, he incorrectly concluded that the area under an arch is very nearly, but not exactly, three times the area of the circle. The first published mathematical demon-

---

*A *cycloid* is the curve traced by a fixed point on the circumference of a circle that rolls, without slipping, along a straight line.

stration that the area is exactly three times that of the generating circle was furnished, in 1644, by his pupil, Evangelista Torricelli (1608–1647), with the use of early integration methods.

Blaise Pascal (1623–1662), when a very young boy, "discovered" that the sum of the angles of a triangle is a straight angle by a simple experiment involving the folding of a paper triangle.

Archimedes (287?–212 B.C.), in his treatise *Method*, has described how he first came to realize, by mechanical considerations, that the volume of a sphere is given by $4\pi r^3/3$, where r is the radius of the sphere. But Archimedes' mathematical conscience would not permit him to accept his mechanical argument as a proof, and he accordingly supplied a rigorous demonstration.

By actually constructing a right circular cone, three times filling it with sand and then emptying the contents into a right circular cylinder of the same radius and height, one would conjecture that the volume of a right circular cone is one-third the product of its altitude and the area of its circular base.

Many first-rate conjectures concerning maxima and minima problems in the calculus of variations were first obtained by soap-film experiments.

One should not deprecate experiments and approaches of this kind, for there is no doubt that much geometry has been "discovered" by such means. Of course, once a geometrical conjecture has been formulated, one must, like Archimedes, establish or refute it by deductive reasoning, and thus completely settle the matter one way or the other. Many a geometrical conjecture has been discarded by the outcome of just one carefully drawn figure or by the examination of some extreme case.

A very fruitful way of making geometrical conjectures is by the employment of analogy, though it must be confessed that many conjectures so made are ultimately proved false. An astonishing amount of space geometry has been discovered via analogy from similar situations in the plane, and in the geometry of higher dimensional spaces analogy has played a very successful role.

There is a pedagogical principle based on the famous law pithily stated by biologists in the form: "Ontogeny recapitulates phylogeny," which simply means that, in general, "the individual repeats the development of the group." The pedagogical principle is that, at least in broad outline, a student should be taught a subject pretty much in the order in which the subject developed over the ages. Take geometry, for example. We have seen that historically geometry progressed through three stages—first subconscious geometry, then scientific geometry, and finally demonstrative geometry. The pedagogical principle claims, then, that geometry should first be presented to young children in its subconscious form, probably through simple art work and simple observations of nature. In this manner the young pupils will subconsciously become aware of a large number of geometrical concepts, such as distance, angle, triangle, quadrilateral, vertical, perpendicular, parallel, straight line, circle, spiral, sphere, cylinder, cone, and so on. Then, somewhat later, this subconscious basis should be evolved into scientific geometry, wherein the pupils induce a considerable array of geometrical facts through experimentation with compasses and straightedge, with ruler and protractor, with scissors and paste, with simple models, and so on. Still later, when the student has become sufficiently sophisticated, geometry can be presented in its demonstrative, or deductive, form, and the advantages and disadvantages of the earlier process can be pointed out.

The weakest part of this geometrical study program in our schools today seems to lie in the second, or scientific, stage of geometry. Not enough time is spent on this stage. There is much to be said for empirical, or experimental, geometry. The time spent here solidifies the students' grasp of many geometrical concepts. It shows them the importance and essential necessity of preliminary inductive procedures in mathematics, at the same time pointing out the shortcomings when the work is not followed up by rigorous demonstrations. What the schoolteachers need in order to make this phase of geometrical learning more extended and more valuable is a good collection of simple but significant geometrical experiments employing inexpensive and easily constructed models. The assembling of a booklet of such experiments is highly recommended to anyone interested in the venture.

Exercises

3.1. The Hindu mathematician Āryabhata the Elder wrote early in the sixth century. His work is a poem of 33 couplets called the *Ganita*. Following are translations of two of the couplets:

The area of a triangle is the product of the altitude and half the base; half of the product of this area and the height is the volume of the solid of six edges.

Half the circumference multiplied by half the diameter gives the area of the circle; this area multiplied by its own square root gives the volume of the sphere.

Show that in each of these couplets Āryabhata is correct in two dimensions but wrong in three. We note that Hindu mathematics remained empirical long after the Greeks had introduced the deductive feature.

3.2. There are two versions of how Thales, when in Egypt, evoked admiration by calculating the height of a pyramid by shadows. The earlier account, given by Hieronymus, a pupil of Aristotle, says that Thales determined the height of the pyramid by measuring the shadow it cast at the moment a man's shadow was equal to his height. The later version, given by Plutarch, says that he set up a stick and then made use of similar triangles. Both versions fail to mention the very real difficulty, in either case, of obtaining the length of the shadow of the pyramid—that is, the distance from the shadow of the apex of the pyramid to the center of the base of the pyramid.

The unaccounted-for difficulty above has given rise to what has become known as the *Thales puzzle*: Devise a method, based on shadow observations and similar triangles and independent of latitude and specific time of day or year, for determining the height of a pyramid. (There is a neat solution employing *two* shadow observations spaced a few hours apart.)

3.3. Assuming the equality of alternate interior angles formed by a transversal cutting a pair of parallel lines, prove the following:

(a) The sum of the angles of a triangle is equal to a straight angle.

(b) The sum of the interior angles of a convex polygon of n sides is equal to $n - 2$ straight angles.

3.4. Assuming the area of a rectangle is given by the product of its two dimensions, establish the following chain of theorems:

(a) The area of a parallelogram is equal to the product of its base and altitude.

(b) The area of a triangle is equal to half the product of any side and the altitude on that side.

(c) The area of a right triangle is equal to half the product of its two legs.

(d) The area of a triangle is equal to half the product of its perimeter and the radius of its inscribed circle.

(e) The area of a trapezoid is equal to the product of its altitude and half the sum of its bases.

(f) The area of a regular polygon is equal to half the product of its perimeter and its apothem.*

(g) The area of a circle is equal to half the product of its circumference and its radius.

3.5. Assuming (1) a central angle of a circle is measured by its intercepted arc, (2) the sum of the angles of a triangle is equal to a straight angle, (3) the base angles of an isosceles triangle are equal, (4) a tangent to a circle is perpendicular to the radius drawn to the point of contact, establish the following chain of theorems:

(a) An exterior angle of a triangle is equal to the sum of the two remote interior angles.

(b) An inscribed angle in a circle is measured by one-half its intercepted arc.

(c) An angle inscribed in a semicircle is a right angle.

(d) An angle formed by two intersecting chords in a circle is measured by one-half the sum of the two intercepted arcs.

(e) An angle formed by two intersecting secants of a circle is measured by one-half the difference of the two intercepted arcs.

(f) An angle formed by a tangent to a circle and a chord through the point of contact is measured by one-half the intercepted arc.

(g) An angle formed by a tangent and an intersecting secant of a circle is measured by one-half the difference of the two intercepted arcs.

---

*The *apothem* of a regular polygon is the perpendicular from the center to any one of its sides.

(h) An angle formed by two intersecting tangents to a circle is measured by one-half the difference of the two intercepted arcs.

3.6. Show empirically, by a simple experiment involving the folding of a paper triangle, that the sum of the angles of a triangle is a straight angle.

3.7. To trisect a central angle AOB of a circle, someone suggests that we trisect the chord AB and then join these points of trisection with O. While this construction may look somewhat reasonable for small angles, show, by taking an angle almost equal to $180°$, that the construction is patently false.

3.8. Two ladders, 60 feet long and 40 feet long, lean from opposite sides across an alley lying between two buildings, the feet of the ladders resting against the bases of the buildings. If the ladders cross each other at a distance of 10 feet above the alley, how wide is the alley?

Find an approximate solution from drawings. An algebraic treatment of this problem requires the solution of a quartic equation. If a and b represent the lengths of the ladders, c the height at which they cross, and x the width of the alley, one can show that

$$(a^2 - x^2)^{-1/2} + (b^2 - x^2)^{-1/2} = c^{-1}.$$

3.9. Let F, V, E denote the number of faces, vertices, and edges of a polyhedron. For the tetrahedron, cube, triangular prism, pentagonal prism, square pyramid, pentagonal pyramid, cube with one corner cut off, cube with a square pyramid erected on one face, we find $V - E + F = 2$. Do you feel that this formula holds for *all* polyhedra?

3.10. There are convex polyhedra all faces of which are triangles (for instance, a tetrahedron), all faces of which are quadrilaterals (for instance, a cube), all faces are pentagons (for instance, a regular dodecahedron). Do you think the list can be continued?

3.11. (a) Consider a convex polyhedron P and let C be any point in its interior. We can imagine a suitable heterogeneous distribution of mass within P such that the center of gravity of P will coincide with C. If this weighted polyhedron should be thrown upon a horizontal floor, it will come to rest on one of its faces (since otherwise we would have perpetual motion). Show that these considerations yield a mechanical argument for the geometrical proposition: "Given a convex polyhedron P and a point C in its interior, then there exists a face F of P such that the foot of the perpendicular from C to the plane of F lies in the interior of F."

(b) Give a geometrical proof of the proposition of part (a).

3.12. Consider an ellipse with semiaxes a and b. If $a = b$ the ellipse becomes a circle and the two expressions

$$P = \pi(a + b) \quad \text{and} \quad P' = 2\pi(ab)^{1/2}$$

each becomes $2\pi a$, which gives the perimeter of the circle. This suggests that P or P' may give the perimeter E of any ellipse. Discuss.

3.13. If the inside of a race track is a noncircular ellipse, and the track is of constant width, is the outside of the track also an ellipse?

3.14. The three altitudes of a triangle are concurrent. Are the four altitudes of a tetrahedron concurrent?

3.15. Formulate theorems in three-space that are analogs of the following theorems in the plane.

(a) The bisectors of the angles of a triangle are concurrent at the center of the inscribed circle of the triangle.

(b) The area of a circle is equal to the area of a triangle the base of which has the same length as the circumference of the circle and the altitude of which is equal to the radius of the circle.

(c) The foot of the altitude of an isosceles triangle is the midpoint of the base of the triangle.

Further Reading

VAN DER WAERDEN, B. L., *Science Awakening*, tr. by Arnold Dresden. New York: Oxford University Press, 1961; New York: John Wiley, 1963 (paperback ed.).

LECTURE **4**

THE FIRST GREAT THEOREM

One of the most attractive, and certainly one of the most famous and most useful, theorems of elementary geometry is the so-called *Pythagorean theorem*, which asserts that "in any right triangle the square on the hypotenuse is equal to the sum of the squares on the two legs." If there is a theorem whose birth merits inclusion as a GREAT MOMENT IN MATHEMATICS, the Pythagorean theorem is probably the prime candidate, for it is perhaps the first truly great theorem in mathematics. But when we come to consider the origin of the theorem, we find ourselves treading on anything but solid ground. Although legend has ascribed the famous theorem to Pythagoras, twentieth-century examination of cuneiform baked clay tablets excavated in Mesopotamia has revealed that the ancient Babylonians of over a thousand years prior to Pythagoras' time were aware of the theorem. Knowledge of the theorem also appears in some ancient Hindu and Chinese works that may go back to the time of Pythagoras, if not earlier. These non-Hellenic and possibly pre-Hellenic references to the theorem, however, contain no proofs of the relationship, and it may well be that Pythagoras, or some member of his renowned fraternity, was the first to furnish a logical demonstration of the theorem.

Let us pause for a moment to say something about Pythagoras and his semimystical brotherhood. Pythagoras is the second person to be mentioned by name in the history of mathematics. Peering through the mythical haze of the past, we gather that Pythagoras was born about 572 B.C. on the Aegean island of Samos, not far from Miletus, the home of the illustrious Thales. Being about fifty years younger than Thales and living so near to him, it may be that Pythagoras studied under the older man. At any rate, he appears, like Thales, to have sojourned at one time in Egypt, and then to have indulged in more extensive travel, probably going as far as India. Returning home after two years of wandering, he found Samos under the tyranny of Polycrates and much of Ionia under Persian dominion, and accordingly he migrated to the Greek seaport of Crotona, located in the boot of southern Italy. There he founded the famous Pythagorean school, which, in addition to being an academy for the study of philosophy, mathematics, and natural science, developed into a closely knit brotherhood with secret rites and observances. In time the political power and aristocratic tendencies of the brotherhood became so great that the democratic forces of southern Italy destroyed the buildings of the school and caused the society to disperse. According to report, Pythagoras fled to Metapontum, where he died, maybe through murder by his pursuers, at the advanced age of 75 or 80. The brotherhood, although scattered, continued to exist for at least two more centuries.

The Pythagorean philosophy, smacking of Hindu origin, rested on the assumption that the whole numbers are the cause of the various qualities of man and matter; in short, the whole numbers rule the universe qualitywise as well as quantitywise. This concept and exaltation of the whole numbers led to their deep study; for, who knows, maybe by unveiling the intricate properties of the whole numbers man might be able, to some degree, to guide or ameliorate his own destiny. Accordingly numbers, and, because of their intimate connection with geometry, geometry too, were assiduously studied. Because the teaching of Pythagoras was entirely oral, and because of the custom of the brotherhood to refer all discoveries back to the revered founder, it is now difficult to know just what mathematical findings should be credited to Pythagoras himself, and which to other members of the fraternity.

Returning to the GREAT MOMENT IN MATHEMATICS under consideration, it is natural to wonder as to the nature of the proof Pythagoras might have given of the great theorem named after him. There has been much conjecture on this, and it is generally felt that the proof was probably a dissection type of proof like the following. Let a, b, c denote the legs and the hypotenuse of the given right triangle, and consider the two squares of Figure 3, each

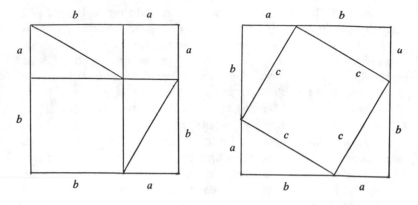

FIG. 3

having $a + b$ as its side. The first square is dissected into six pieces, namely, the two squares on the legs and four right triangles congruent to the given triangle. The second square is dissected into five pieces, namely, the square on the hypotenuse and again four right triangles congruent to the given triangle. By subtracting equals from equals, it now follows that the square on the hypotenuse is equal to the sum of the squares on the legs.

To prove that the central piece of the second dissection is actually a square of side c, we need to employ the fact that the sum of the angles of a right triangle is equal to two right angles. But this fact for the general triangle has been attributed to the Pythagoreans. Since a proof of this general fact requires, in turn, a knowledge of some properties of parallels, the early Pythagoreans are also credited with the development of that theory.

Perhaps no theorem in all of mathematics has received more diverse proofs than has the Pythagorean theorem. In the second edition of his book, *The Pythagorean Proposition*,* E. S. Loomis has collected and classified 370 demonstrations of this famous theorem.

---

*Ann Arbor, Mich.: private printing, Edward Brothers, 1940. Reprint available from The National Council of Teachers of Mathematics, Washington, D.C.

Two areas, or two volumes, P and Q, are said to be *congruent by addition* if they can be dissected into corresponding pairs of congruent pieces. They are said to be *congruent by subtraction* if corresponding pairs of congruent pieces can be adjoined to P and Q to give two new figures that are congruent by addition. There are many proofs of the Pythagorean theorem which achieve their end by showing that the square on the hypotenuse of the right triangle is congruent either by addition or subtraction to the combined squares on the legs of the right triangle. The proof sketched above, and thought perhaps due to Pythagoras, is a congruency-by-subtraction proof.

Figures 4 and 5 suggest two congruency-by-addition proofs of the Pythagorean theorem, the first given by H. Perigal in 1873* and the second by H. E. Dudeney in 1917. Figure 6 suggests a congruency-by-subtraction proof said to have been devised by Leonardo da Vinci (1452–1519).

It is interesting that any two equal polygonal areas are congruent by addition, and the dissection can always be effected with straight-edge and compasses. On the other hand, in 1901, Max Dehn showed that two equal polyhedral volumes are not necessarily congruent by either addition or subtraction. In particular, it is impossible to dissect a regular tetrahedron into polyhedral pieces that can be reassembled to form a cube. Euclid, in his *Elements* (ca. 300 B.C.), occasionally employs dissection methods to establish equivalence of areas.

The elegant proof of the Pythagorean theorem given by Euclid in Proposition 47 of Book I of his *Elements* depends upon the diagram of Figure 7, sometimes referred to as the Franciscan's cowl, or as the bride's chair. A précis of the proof runs as follows: $(AC)^2 = 2 \triangle JAB = 2 \triangle CAD = ADKL$. Similarly, $(BC)^2 = BEKL$. Therefore $(AC)^2 + (BC)^2 = ADKL + BEKL = (AB)^2$.

High school teachers sometimes show their students the curious proof of the Pythagorean theorem given by the Hindu mathematician and astronomer Bhāskara, who flourished around 1150. It is a dissection proof in which the square on the hypotenuse is cut up, as indicated in Figure 8, into four triangles, each congruent to

---

*This was a rediscovery, for the dissection was known to Tâbit ibn Qorra (826–901).

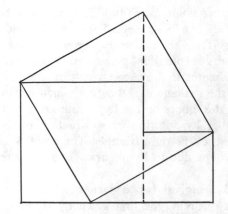

FIG. 4

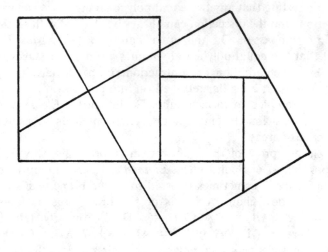

FIG. 5

the given right triangle, plus a square with side equal to the difference of the legs of the given right triangle. The pieces are easily
rearranged to give the sum of the squares on the two legs. Bhāskara
drew the figure and offered no further explanation than the word

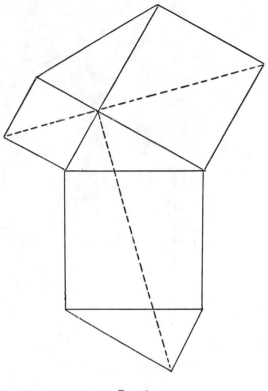

FIG. 6

"Behold!" Of course, a little algebra supplies a proof. For, if c is the hypotenuse and a and b are the legs of the given right triangle,

$$c^2 = 4(ab/2) + (b - a)^2 = a^2 + b^2.$$

Perhaps a better "behold" proof of the Pythagorean theorem would be a dynamical one on movie film wherein the square on the hypotenuse is continuously transformed into the sum of the squares on the legs by passing through the stages indicated in Figure 9.

Bhāskara also gave a second demonstration of the Pythagorean

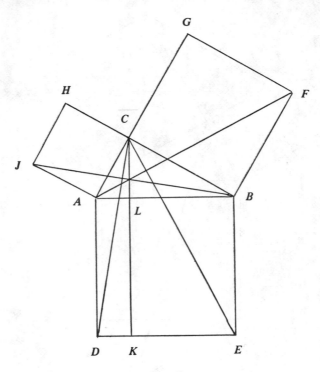

Fig. 7

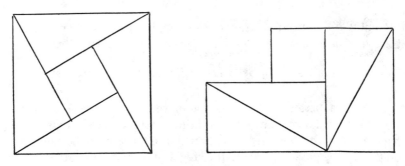

Fig. 8

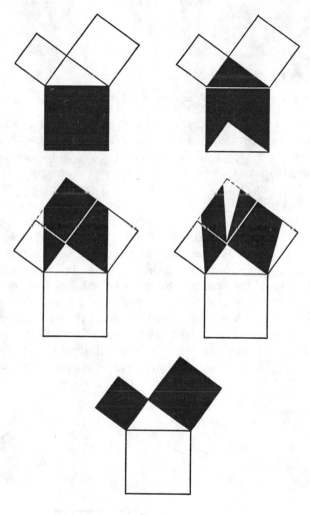

Fig. 9

theorem by drawing the altitude on the hypotenuse. From similar right triangles in Figure 10 we have

$$c/b = b/m \text{ and } c/a = a/n,$$

or

$$cm = b^2 \text{ and } cn = a^2.$$

Adding, we get

$$a^2 + b^2 = c(m + n) = c^2.$$

This proof was rediscovered in the seventeenth century by the English mathematician John Wallis (1616–1703).

A few of our country's presidents have been tenuously connected with mathematics. George Washington was a noted surveyor, Thomas Jefferson did much to encourage the teaching of higher mathematics in the United States, and Abraham Lincoln is said to have learned logic by studying Euclid's *Elements*. More creative was James Abram Garfield (1831–1881), the country's twentieth president, who in his student days developed a keen interest and fair ability in elementary mathematics. It was in 1876, while he was a member of the House of Representatives, and five years before he became President of the United States, that he independently discovered a very pretty proof of the Pythagorean theorem. He hit upon the proof in a mathematics discussion with some other members of Congress, and the proof was subsequently printed up in the *New England Journal of Education*. Students of high school

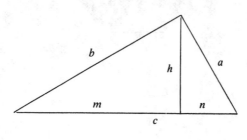

Fig. 10

geometry are always interested to see the proof, which can be presented immediately after the formula for the area of a trapezoid has been covered. The proof depends upon calculating the area of the trapezoid of Figure 11 in two different ways—first by the formula for the area of a trapezoid (as the product of half the sum of the parallel sides and the perpendicular distance between these sides) and then as the sum of three right triangles into which the trapezoid can be dissected. Equating the two expressions so found for the area of the trapezoid, we find (see Figure 11)

$$(a + b)(a + b)/2 - 2[(ab)/2] + c^2/2$$

or

$$a^2 + 2ab + b^2 = 2ab + c^2,$$

whence

$$a^2 + b^2 = c^2.$$

Since a trapezoid, as pictured, exists for any right triangle of legs a and b and hypotenuse c, the Pythagorean theorem has been established.

Like many other great theorems, the Pythagorean theorem has received numerous extensions. Even in Euclid's time certain generalizations of the theorem were known. For example, Proposition 31 of Book VI of the *Elements* states: *In a right triangle the area of a figure described on the hypotenuse is equal to the sum of the areas of similar figures similarly described on the two legs.* This generalization merely replaces the three squares on the three sides of the right triangle by any three similar and similarly described figures. A more worthy generalization stems from Propositions 12 and 13 of Book II. A combined and somewhat modernized statement of these two propositions is: *In a triangle, the square of the side opposite an obtuse (acute) angle is equal to the sum of the squares on the other two sides increased (decreased) by twice the product of one of these sides and the projection of the other side on it.* That is, in the notation of Figure 12.

$$(AB)^2 = (BC)^2 + (CA)^2 \pm 2(BC)(DC),$$

the plus or minus sign being taken according as angle C of triangle *ABC* is obtuse or acute. If we employ directed line segments we may combine Propositions 12 and 13 of Book II and Proposition 47 of Book I (the Pythagorean theorem) into the single statement: *If in triangle ABC, D is the foot of the altitude on side BC, then*

$$(AB)^2 = (BC)^2 + (CA)^2 - 2(BC)(DC).$$

Since $DC = CA \cos BCA$, we recognize this last statement as essentially the so-called *law of cosines*, and the law of cosines is indeed a fine generalization of the Pythagorean theorem.

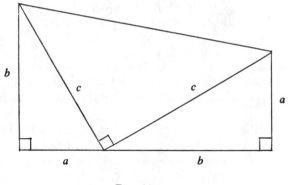

Fig. 11

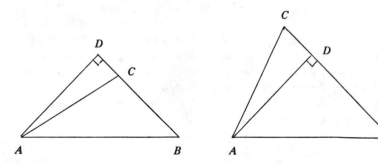

Fig. 12

But perhaps the most remarkable extension of the Pythagorean theorem that dates back to the days of Greek antiquity is that given by Pappus of Alexandria (ca. A.D. 300) at the start of Book IV of his *Mathematical Collection*. The Pappus extension of the Pythagorean theorem is as follows (see Figure 13): *Let ABC be any triangle and CADE, CBFG any parallelograms described externally on sides CA and CB. Let DE and FG meet in H and draw AL and BM equal and parallel to HC. Then the area of parallelogram ABML is equal to the sum of the areas of parallelograms CADE and CBFG.* The proof is easy, for we have $CADE = CAUH = SLAR$ and $CBFG = CBVH = SMBR$. Hence $CADE + CBFG = SLAR + SMBR = ABML$. It is to be noted that the Pythagorean theorem has been generalized in two directions, for the right triangle in the Pythagorean theorem has been replaced by *any* triangle, and the squares on the legs of the right triangle have been replaced by *any* parallelograms.

The student of high school geometry can hardly fail to be interested in the Pappus extension of the Pythagorean theorem, and the proof of the extension can serve as a nice exercise for the student. Perhaps the more gifted student of geometry might like to try his hand at establishing the further extension (to three-space) of the Pappus extension: *Let ABCD (see Figure 14) be any tetrahedron and let ABD-EFG, BCD-HIJ, CAD-KLM be any three triangular prisms described externally on the faces ABD, BCD, CAD of ABCD. Let Q be the point of intersection of the planes EFG, HIJ, KLM, and let ABC-NOP be the triangular prism whose edges AN, BO, CP are translates of the vector QD. Then the volume of ABC-NOP is equal to the sum of the volumes of ABD-EFG, BCD-HIJ, CAD-KLM.* A proof analogous to the one given above for the Pappus extension can be supplied.

We give finally, without proof, a three-space analogue of the Pythagorean theorem that is often referred to as *de Gua's theorem.* *

---

*Named after J. P. de Gua de Malves (1712-1785), who presented the proposition to the Paris Academy of Sciences in 1783. The theorem, however, had been known to Descartes (1596-1650) and his contemporary J. Faulhaber (1580-1635). It is a special case of a more general theorem that Tinseau had presented to the Paris Academy of Sciences in 1774.

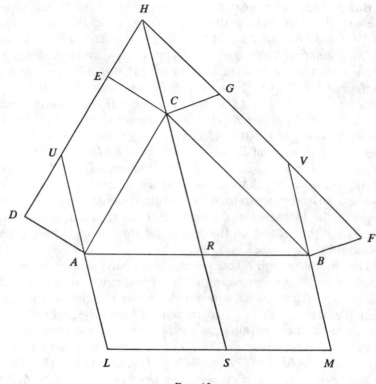

Fig. 13

We first formulate some definitions. A tetrahedron having a trihedral angle all face angles of which are right angles is called a *trirectangular tetrahedron,* and the trihedral angle is called the *right angle* of the tetrahedron. The face opposite the right angle is called the *base* of the tetrahedron. De Gua's theorem may now be stated as follows: *The square of the area of the base of a trirectangular tetrahedron is equal to the sum of the squares of the areas of its other three faces.* We leave the matter of proof to any enterprising reader.

With the mounting interest in space exploration and the possibility of life in other parts of the universe, suggestions have appeared from time to time concerning the construction on the earth

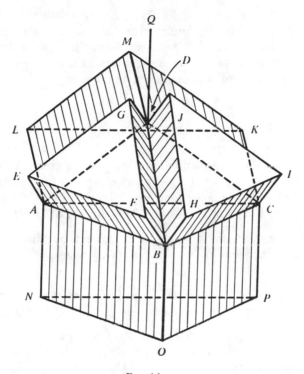

FIG. 14

of some enormous device that would indicate to possible outside observers that there is intelligence on our planet. The most favored device seems to be a mammoth configuration illustrating the Pythagorean theorem, built on the Sahara Desert, the Steppes of Russia, or some other vast area. All intelligent beings must be acquainted with this remarkable and certainly nontrivial theorem of Euclidean geometry, and it does seem difficult to think of a better visual device for the purpose under consideration.

In 1971 Nicaragua issued a series of stamps paying homage to the world's "ten most important mathematical formulas." Each stamp features a particular formula accompanied by an appropriate illustration and carries on its reverse side a brief statement in Spanish concerning the importance of the formula. One of the stamps in the series honors the Pythagorean relation "$a^2 + b^2 = c^2$." It must be pleasing to scientists and mathematicians to see these formulas so honored, for these formulas have certainly contributed far more to human development than did many of the kings and generals so often featured on stamps.

Exercises

4.1. Prove that two parallelograms having a common base and equal altitudes have equal areas by showing them to be either congruent by addition or congruent by subtraction. (This is the method employed by Euclid in Proposition 35 of Book I of his *Elements*.)

4.2. Show that any triangle is congruent by addition to the equivalent rectangle having for length a longest side of the triangle.

4.3. Fill in the details of the dissection proof of the Pythagorean theorem thought perhaps to have been given by Pythagoras.

4.4. Complete the details in the dissection proof of the Pythagorean theorem credited to: (a) H. Perigal, (b) H. E. Dudeney, (c) Leonardo da Vinci.

4.5. (a) There are reports that ancient Egyptian surveyors laid out right angles by constructing 3-4-5 triangles with a rope divided into 12 equal parts by 11 knots. Show how this can be done.

(b) Since there is no documentary evidence to the effect that the Egyptians were aware of even a particular case of the Pythagorean theorem, the following purely academic problem arises: Show, without using the Pythagorean theorem, its converse, or any of its consequences, that the 3-4-5 triangle is a right triangle. Solve this problem by means of Figure 15, which appears in the *Chóu-peï*, the oldest known Chinese mathematical work, which may date back to the second millennium B.C.

4.6. Supply a proof of the three-space analogue of the Pappus extension of the Pythagorean theorem.

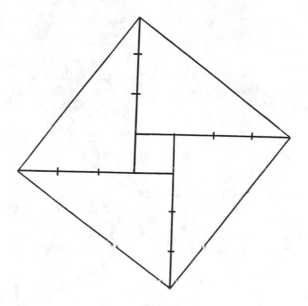

FIG. 15

4.7. The edges issuing from the right angle of a trirectangular tetrahedron are called the *legs* of the tetrahedron, and the perpendicular from the vertex of the right angle to the base is called the *altitude* of the tetrahedron.

(a) Prove that the sum of the squares of the reciprocals of the legs of a trirectangular tetrahedron is equal to the square of the reciprocal of the altitude of the tetrahedron.

(b) Prove de Gua's theorem.

4.8. Establish the following generalization of the Pythagorean theorem given by Tâbit ibn Qorra: If triangle ABC is any triangle, and if B' and C' are points on BC such that $\angle AB'B = \angle AC'C = \angle A$, then $(AB)^2 + (AC)^2 = BC(BB' + CC')$.

Show that when $\angle A$ is a right angle this theorem becomes the Pythagorean theorem.

4.9. What is the Pythagorean relation for a right spherical triangle of legs a and b and hypotenuse c, where a, b, and c are angular measurements?

4.10. State and prove the converse of the Pythagorean theorem. (This is Proposition 48, the final proposition, of Book I of Euclid's *Elements*.)

Further Reading

BOLTYANSKII, *Equivalent and Equidecomposable Figures*, tr. by A. K. Henn and C. E. Watts. Boston: D. C. Heath, 1963.

HEATH, T. L., *History of Greek Mathematics*, 2 vols. New York: Oxford University Press, 1931.

LOOMIS, E. S., *The Pythagorean Proposition*, 2nd ed. Ann Arbor, Mich.: privately printed, Edwards Brothers, 1940.

PRECIPITATION OF THE FIRST CRISIS

The first numbers we encounter as we grow up from early childhood are the so-called *natural numbers, or positive integers*: 1, 2, 3, ... These numbers are abstractions that arise from the process of counting finite collections of objects. Somewhat later we realize that the needs of daily life require us, in addition to counting individual objects, to measure various quantities, such as length, weight, and time. To satisfy these simple measuring needs, *fractions* are required, for seldom will a length, to take an example, appear to contain an exact integral number of a prechosen linear unit. For some measurements, such as recording very low temperatures, the *zero* and *negative integers* and the *negative fractions* are found convenient. Our number system has been widened. But, if we define a *rational number* as the quotient of two integers, p/q, $q \neq 0$, then this system of rational numbers, since it contains all the integers and all the fractions, is quite sufficient for all our practical measuring purposes.

Now the rational numbers have a simple geometrical representation. Mark two distinct points O and I (see Figure 16) on a horizontal straight line, I to the right of O, and choose the segment OI as a unit of length. If we let O and I represent the numbers 0 and 1, respectively, then the positive and negative integers can be represented by a set of points on the line spaced at unit intervals apart, the positive integers being represented to the right of O and the negative integers to the left of O. The fractions with denominator q may then be represented by the points that divide each of the unit intervals into q equal parts. For each rational number, then, there is a unique point on the line. To the early mathematicians it seemed evident, as indeed it seems to anyone today who

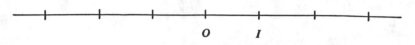

O I

FIG. 16

has not yet been initiated into the deeper mysteries of the number line, that all the points on the line are in this way used up; ordinary common sense seems to indicate this to us.

It must have been a genuine mental shock for man to learn that there are points on the number line not corresponding to any rational number. This discovery was certainly one of the greatest achievements of the early Greeks, and it seems to have occurred some time in the fifth or sixth century B.C. among the ranks of the Pythagorean brotherhood. A truly GREAT MOMENT IN MATHEMATICS had arisen.

In particular, the Pythagoreans found that there is no rational number corresponding to the point P on the number line (see Figure 17) where the distance OP is equal to the diagonal of a square having a unit side. Later, many other points on the number line were found not corresponding to any rational number. New numbers had to be invented to correspond to such points, and since these numbers cannot be rational numbers (that is, *ratio* numbers), they came to be called *irrational numbers*.

Since, by the Pythagorean theorem, the length of a diagonal of a square of unit side is $\sqrt{2}$, in order to prove that the point P above cannot be represented by a rational number, it suffices to show that $\sqrt{2}$ is irrational. To this end, we first observe that, for a positive integer s, s^2 is even if and only if s is even. Now suppose, for the purpose of argument, that $\sqrt{2}$ is rational, that is, that $\sqrt{2} = p/q$, where p and q are relatively prime integers.* Then

$$p = q\sqrt{2},$$

or

$$p^2 = 2q^2.$$

---

*Two integers are *relatively prime* if they have no common positive integral factor other than unity. Thus 5 and 18 are relatively prime, whereas 12 and 18 are not relatively prime.

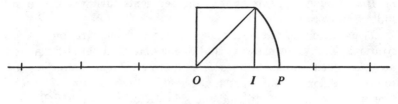

FIG. 17

Since p^2 is twice an integer, we see that p^2, and hence p, must be even. Put $p = 2r$. Then the last equation becomes

$$4r^2 = 2q^2,$$

or

$$2r^2 = q^2,$$

from which we conclude that q^2, and hence q, must be even. But this is impossible since p and q were assumed to be relatively prime. Thus the assumption that $\sqrt{2}$ is rational has led to an impossible situation, and the assumption must be abandoned.

This proof of the irrationality of $\sqrt{2}$ is essentially the traditional one reported by Aristotle (384–322 B.C.). According to Plato (427–347 B.C.), after $\sqrt{2}$ had been shown to be irrational, Theodorus of Cyrene (ca. 425 B.C.) showed that $\sqrt{3}$, $\sqrt{5}$, $\sqrt{6}$, $\sqrt{7}$, $\sqrt{8}$, $\sqrt{10}$, $\sqrt{11}$, $\sqrt{12}$, $\sqrt{13}$, $\sqrt{14}$, $\sqrt{15}$, $\sqrt{17}$ are also irrational.

The discovery of the existence of irrational numbers upset another intuitive belief held by the early Greeks. Given any two line segments, common sense seemed to dictate that there must be some third line segment, perhaps very, very small, that can be marked off a whole number of times into each of the two given segments. Indeed, almost anyone today who has not yet learned otherwise intuitively feels the same way. But take as the two segments a side s and a diagonal d of a square. Now if there exists a third segment t that can be marked off a whole number of times into s and d we would have $s = qt$ and $d = pt$, where p and q are positive integers. But $d = s\sqrt{2}$, whence $pt = qt\sqrt{2}$. That is, $p = q\sqrt{2}$, or $\sqrt{2} = p/q$, a rational number. Contrary to intuition, then, there exist *incom-*

mensurable line segments, that is, line segments having no common unit of measure.

Let us sketch an alternative, geometrical, demonstration of the irrationality of $\sqrt{2}$ by showing that a side and diagonal of a square are incommensurable. Suppose the contrary. Then, according to this supposition, there exists a segment AP (see Figure 18) such that both the diagonal AC and the side AB of a square $ABCD$ are integral multiples of AP; that is, AC and AB are commensurable with respect to AP. On AC lay off $CB_1 = AB$ and draw B_1C_1 perpendicular to CA. One may easily prove that $C_1B = C_1B_1 = AB_1$. Then $AC_1 = AB - AB_1$ and AB_1 are commensurable with respect to AP. But AC_1 and AB_1 are a diagonal and a side of a square of dimensions less than half those of the original square. It follows that by repeating the process enough times we may finally obtain a square whose diagonal AC_n and side AB_n are commensurable with respect to AP, and $AC_n < AP$. This absurdity proves the theorem.

One notes that each of the above proofs of the irrationality of $\sqrt{2}$ employs the indirect, or *reductio ad absurdum,* method of proof. The eminent English mathematician G. H. Hardy (1877–1947) has made a delightful remark about this type of proof. In the game of chess a *gambit* is any maneuver in which a pawn or a piece is sacrificed in order to obtain an advantageous attack. Hardy pointed out that *reductio ad absurdum* "is a far finer gambit than any chess gambit: a chess player may offer the sacrifice of a pawn or a piece, but a mathematician offers *the game.*"* Reductio ad absurdum* emerges as the most stupendous gambit conceivable.

An interesting encounter with an irrational number arose in ancient times when Greek geometers tried to construct a regular polygon of five sides. They had easily succeeded in constructing regular polygons of three and four sides, namely, an equilateral triangle and a square, and, of course, the construction of a regular polygon of six sides presented no difficulty. But the construction of a regular polygon of five sides—that is, a regular pentagon—is quite another matter. Success would be assured if one can construct an angle of 36°, inasmuch as twice this angle, or 72°, is the

---

*G. H. Hardy, *A Mathematician's Apology.* New York: Cambridge University Press, 1941, p. 34.

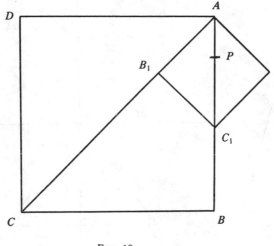

FIG. 18

central angle subtended by one side of a regular pentagon inscribed in a circle. Since, in an isosceles triangle each of whose base angles is twice the vertex angle of the triangle (see Figure 19), the base angles are 72° and the vertex angle is 36°, the problem is reduced to the construction of such an isosceles triangle. Let AC in Figure 19 bisect the base angle OAB. Then $OC = AC = AB$ and triangle BAC is similar to triangle AOB. Taking $OA = 1$ and setting $AB = x$, we then have, in turn,

$$AB/BC = OA/AB, \ x/(1 - x) = 1/x, \ x^2 + x - 1 = 0.$$

It follows that $x = (\sqrt{5} - 1)/2$. The construction of this x is an easy matter, and is indicated in Figure 20, where $OA = 1$ and $MO = 1/2$, and, consequently, $AM = \sqrt{5}/2$ and

$$AB = AN = AM - MN = (\sqrt{5} - 1)/2 = x.$$

The construction of the inscribed regular pentagon now easily follows.

When a line segment OB (like OB in Figure 19) is divided by a point C such that the longer segment OC is a mean proportional

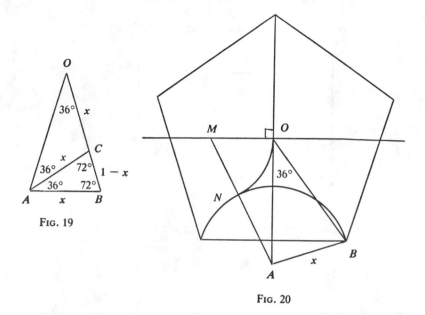

FIG. 19

FIG. 20

between the shorter segment CB and the whole segment OB, that is, when

$$CB/OC = OC/OB,$$

the Greeks said that the line segment OB is divided into *golden section*. We found above that if x represents either of the ratios CB/OC or OC/OB, then $x = (\sqrt{5} - 1)/2$. This number, or sometimes its reciprocal

$$y = 1/x = (\sqrt{5} + 1)/2 \doteq 1.618,$$

is called the *golden ratio*, and this ratio seems to occur ubiquitously in nature and elsewhere.

We shall comment on the occurrences of the golden ratio in nature later on, in LECTURE 15. We here remark that psychological tests tend to show that to most people the rectangle that appears most pleasing to the eye is the one whose ratio of width to length is the golden ratio x. This rectangle, which is called the *golden rectangle*, is fundamental in an art technique known as "dynamic symmetry," which has been intensively studied by Jay Hambidge and others. The golden ratio and the golden rectangle have been observed in Greek architecture and Greek pottery, and have been applied to sculpture, painting, architectural design, furniture design, and type display. A number of artists, such as the well-known American painter George Bellows, have extensively used the principles of dynamic symmetry in their work.

A fundamental difference between rational and irrational numbers became manifest after the invention of decimal fractions. It is easily shown that any rational number possesses either a terminating or a repeating decimal expansion, and conversely, any terminating or repeating decimal expansion represents a rational number. For example: $7/4 = 1.75$, $47/22 = 2.13\overline{63}$, where the bar over the 63 means that the decimal segment 63 is endlessly repeated. It follows that the decimal expansion of an irrational number is nonterminating and nonrepeating, and conversely, any nonterminating and nonrepeating decimal expansion represents some irrational number.

The distinction between the decimal expansions of rational and irrational numbers is very useful in establishing certain properties of these numbers. Suppose, for example, we wish to show that there exists a rational number between any two distinct positive irrational numbers. Denote the two irrational numbers by a and b, $0 < a < b$, and let their decimal expansions be

$$a = a_0.a_1a_2 \ldots \quad \text{and} \quad b = b_0.b_1b_2 \ldots .$$

Let i be the first value of n for which $a_n \neq b_n$ ($n = 0, 1, 2, \ldots$). Then

$$c = b_0.b_1b_2 \ldots b_i$$

is a rational number between a and b.

A real number is called *simply normal* if all ten digits occur with equal frequency in its decimal representation, and it is called *normal* if all blocks of digits of the same length occur with equal frequency. It is believed, but not known, that π, e, and $\sqrt{2}$, for example, are normal numbers. To obtain statistical evidence of the supposed normalcy of the above numbers, their decimal expansions have been carried out to great numbers of decimal places.

In 1967, British mathematicians, working with a computer, carried the decimal expansion of $\sqrt{2}$ to 100,000 places. In 1971, Jacques Dutka, of Columbia University, found $\sqrt{2}$ to over one million places—after 47.5 hours of computer time, the electronic machine ticked off the decimal expansion of $\sqrt{2}$ to at least 1,000,082 correct places, filling 200 pages of tightly spaced computer print-out, each page containing 5000 digits. This is the longest approximation to an irrational number ever computed.

Exercises

5.1. (a) Fill in the details of the geometric proof of the irrationality of $\sqrt{2}$ sketched in the lecture text.

(b) Draw a 60°-30° right triangle; mark off the longer leg, from the 30° angle vertex, on the hypotenuse; draw a perpendicular to the hypotenuse from the dividing point. Using this figure, formulate a geometrical proof of the irrationality of $\sqrt{3}$.

5.2. (a) Prove that the straight line through the points (0, 0) and (1, $\sqrt{2}$) passes through no point, other than (0, 0), of the coordinate lattice.

(b) Show how the coordinate lattice may be used for finding rational approximations of $\sqrt{2}$.

5.3. If p is a prime number and n an integer greater than 1, show that $\sqrt[n]{p}$ is irrational.

5.4. (a) Show that $\log_{10}2$ is irrational.

(b) Generalize part (a) by showing that $\log_a b$ is irrational if a and b are positive integers, $a > 1$, and one of them contains a prime factor not contained in the other.

5.5. (a) Show that the sum of a rational and an irrational number is an irrational number.

(b) Show that the product of a rational and an irrational number is an irrational number.

5.6. (a) The symbol of the Pythagorean brotherhood was the *pentagram,* or five-pointed star formed by the five diagonals of a regular pentagon. Prove that each of the five sides of a pentagram divides into golden section the two sides of the pentagram that it intersects.

(b) Let point G divide line segment AB in golden section, where AG is the longer segment. On AB mark off $AH = GB$. Show that H divides AG in golden section.

(c) Show that if a square is cut off one end of a golden rectangle, the remaining piece is a golden rectangle.

(d) Show that 5/8 overestimates the golden ratio $x = (\sqrt{5} - 1)/2$, with an error which is less than 3 percent.

(e) If x is the golden ratio $(\sqrt{5} - 1)/2$, show that

$$x = \frac{1}{1 + x} = \cfrac{1}{1 + \cfrac{1}{1 + \cfrac{1}{1 + x}}} = \text{etc.}$$

5.7. (a) Construct, with straightedge and compasses, a regular pentagon given one side of the pentagon.

(b) Construct, with straightedge and compasses, a regular pentagon given one diagonal of the pentagon.

(c) Construct, with straightedge and compasses, a regular polygon of 15 sides.

(d) Suppose r and s are relatively prime positive integers and that a regular r-gon and a regular s-gon are constructible with straightedge and compasses. Show that a regular rs-gon is also so constructible.

(e) Establish Proposition XIII, 10, of Euclid's *Elements: A side of a regular pentagon, of a regular hexagon, and of a regular decagon inscribed in the same circle constitute the sides of a right triangle.*

5.8 (a) Prove that the decimal expansion of a rational number is either terminating or repeating.

(b) Prove that a terminating or repeating decimal expansion represents some rational number.

(c) Find the rational number having $3.2\overline{39}$ for its decimal expansion.

5.9. (a) Show that

$$0.101001000100001 \ldots ,$$

where the number of 0's between successive 1's increases each time by one, is an irrational number.

(b) Show that

$$0.12345678910111213 \ldots ,$$

in which the decimal expansion consists of the successive positive integers, is an irrational number.

5.10. (a) Prove that between any two distinct rational numbers there are infinitely many rational numbers.

(b) Prove that between any two distinct rational numbers there are infinitely many irrational numbers.

(c) Prove that between any two distinct irrational numbers there are infinitely many rational numbers.

(d) Prove that between any two distinct irrational numbers there are infinitely many irrational numbers.

Further Reading

HAMBIDGE, JAY, *The Elements of Dynamic Symmetry.* New York: Dover Publications, 1967.

HEATH, T. L., *History of Greek Mathematics*, 2 vols. New York: Oxford University Press, 1931.

HUNTLEY, H. E., *The Divine Proportion, a Study in Mathematical Beauty.* New York: Dover Publications, 1970.

RESOLUTION OF THE FIRST CRISIS

The discovery of irrational numbers and of incommensurable magnitudes caused considerable consternation in the Pythagorean ranks. First of all, it seemed to deal a mortal blow to the Pythagorean philosophy that all depends upon the whole numbers— after all, how does an irrational number, like $\sqrt{2}$, depend on the whole numbers if it cannot be written as the ratio of two such numbers? Next, it seemed contrary to common sense, for it was strongly felt intuitively that any magnitude could be expressed by *some* rational number. The geometric counterpart was equally startling, for, again contrary to intuition, there exist line segments having no common unit of measure. But the whole Pythagorean theory of proportion and similar figures was built upon the seemingly obvious assumption that any two line segments are commensurable, that is, do have some common unit of measure. A large portion of geometry that the Pythagoreans had felt was established suddenly had to be scrapped as unsound because the proofs were invalid. A serious crisis in the foundations of mathematics was precipitated. So great was the "logical scandal" that, according to report, efforts were made for a while to keep the matter secret, and one legend has it that the Pythagorean Hippasus of Metapontum perished at sea for his impiety in disclosing the secret to outsiders, or (according to another version) was banished from the Pythagorean community and a tomb erected for him as though he were dead.

Let us see, by way of an example, how the early Pythagoreans believed they had established a basic proposition concerning areas of triangles.

THEOREM. *The areas of two triangles having the same altitude are to one another as their bases.*

53

Early Pythagorean proof. Let the triangles be ABC and ADE, the bases BC and DE lying on the same straight line MN, as in Figure 21. We wish to prove that

$$\triangle ABC : \triangle ADE = BC : DE.$$

Let (according to the early Pythagorean belief that any two line segments are commensurable) some common unit of measure of BC and DE be contained, say, p times in BC and q times in DE. Mark off these points of division on BC and DE and connect them with vertex A. Then triangles ABC and ADE are divided, respectively, into p and q smaller triangles, all having a common altitude and equal bases and, therefore, by an earlier established result, the same area. It follows that

$$\triangle ABC : \triangle ADE = p : q = BC : DE,$$

and the proposition is established.

With the later discovery that two line segments need not be commensurable, this proof, along with many others, became inadequate, and the very disturbing "logical scandal" came into existence.

This "logical scandal" was the first known crisis in the foundations of mathematics, and it was neither easily nor quickly resolved. Finally, about 370 B.C., the brilliant Greek mathematician Eudoxus, a pupil of Plato and of the gifted Pythagorean, Archytas, cleverly resolved the "scandal" by formulating a definition of proportion, or equality of two ratios, that is entirely independent of the commensurability or incommensurability of the magnitudes involved. This definition, which marks a GREAT MOMENT IN MATHEMATICS, runs as follows:

Magnitudes are said to be in the same ratio, the first to the second and the third to the fourth, when, if any equimultiples whatever be taken of the first and third, and any equimultiples whatever of the second and fourth, the former equimultiples alike exceed, are alike equal to, or are alike less than the latter equimultiples taken in corresponding order.

This verbose definition, which appears more complicated than it really is, simply means that if A, B, C, D are any four unsigned

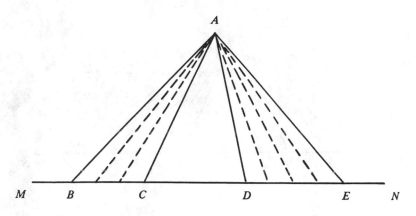

FIG. 21

magnitudes, A and B being of the same kind (both line segments, or angles, or areas, or volumes) and C and D being of the same kind, then the ratio of A to B is equal to that of C to D if for arbitrary positive integers m and n,

$$mA \gtreqless nB \text{ according as } mC \gtreqless nD.$$

Let us now apply the Eudoxian definition of proportion by proving anew the proposition considered above.

THEOREM. *The areas of two triangles having the same altitude are to one another as their bases.*

Eudoxian proof. On CB produced (see Figure 22), mark off, successively from B, $m - 1$ segments equal to CB, and connect the points of division, $B_2, B_3, \ldots, B_m$, with vertex A. Similarly, on DE produced, mark off, successively from E, $n - 1$ segments equal to DE, and connect the points of division, $E_2, E_3, \ldots E_n$, with vertex A. Then

$$B_m C = m(BC), \qquad \triangle AB_m C = m(\triangle ABC),$$

$$DE_n = n(DE), \qquad \triangle ADE_n = n(\triangle ADE).$$

Also, by earlier established results,

$$\triangle AB_m C \gtreqless \triangle ADE_n \text{ according as } B_m C \gtreqless DE_n.$$

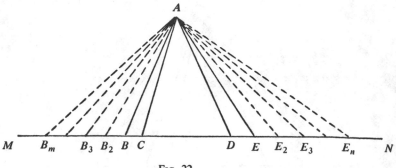

FIG. 22

That is

$$m(\triangle ABC) \gtreqless n(\triangle ADE) \quad \text{according as} \quad m(BC) \gtreqless n(DE),$$

whence (by the Eudoxian definition of proportion)

$$\triangle ABC : \triangle ADE = BC : DE,$$

and the proposition is established.

In the above proof, no mention was made of commensurable and incommensurable quantities, since the Eudoxian definition applies equally to both situations. The "logical scandal" of the earlier Pythagorean treatment has been ingeniously bypassed.

An elegant exposition of the Eudoxian theory of proportion was later, about 300 B.C., presented by Euclid in Book V of his famous *Elements*. Bernhard Bolzano (1781–1848), who in his day was a frowned-upon Czechoslovakian priest and an overlooked mathematician, has told a charming little anecdote about himself in which Euclid and his Book V played the role of a physician. Bolzano was on vacation in Prague when he was attacked by an illness that manifested itself in bodily chills and painful weariness. To take his mind from his condition, he picked up Euclid's *Elements* and for the first time read the masterly exposition of the Eudoxian doctrine of ratio and proportion set out in Book V. The ingenuity of the treatment filled him with such vivid pleasure that, he said, he completely recovered from his illness. Ever after, when any of his friends felt indisposed, he recommended as a cure the reading of Euclid's presentation of the Eudoxian theory.

The pedagogical difficulties inherent in the incommensurable case of certain propositions of plane and solid geometry made themselves felt right up into present times. Viewing the Eudoxian procedure as too abstract for beginning students, writers of many of the high school textbooks advocated proofs involving two cases, the commensurable case and the incommensurable case, the commensurable case to be handled in the early Pythagorean fashion and simple limit notions to be used in dealing with the incommensurable case. Sometimes the incommensurable case was relegated to an appendix, to be covered at the instructor's discretion, and sometimes it was omitted entirely as being beyond the rigor of the course. In most recent high school geometry texts, the incommensurable case is taken care of by extended assumption, such as knowledge of the completed real number system and its basic properties.

Let us return to the proposition considered earlier and attack the incommensurable case with the aid of some simple limit notions. In particular, we will employ the following easily accepted fundamental theorem of limit theory.

FUNDAMENTAL LIMIT THEOREM. *If two variables are always equal, and each approaches a limit, then these limits are equal.*

Now for the troublesome incommensurable case under consideration.

THEOREM. *The areas of two triangles having the same altitude are to one another as their bases.*

Incommensurable case. Suppose, in Figure 23, that BC and DE are incommensurable. We wish to show that

$$\triangle ABC : \triangle ADE = BC : DE.$$

Divide BC into n equal parts, BR being one of the parts. On DE mark off a succession of segments equal to BR, finally arriving at a point F on DE such that $FE < BR$. By the commensurable case (assumed already established in early Pythagorean fashion),

$$\triangle ABC : \triangle ADF = BC : DF.$$

Now let $n \to \infty$. Then $DF \to DE$ and $\triangle ADF \to \triangle ADE$, whence

$$\triangle ABC / \triangle ADF \to \triangle ABC / \triangle ADE \quad \text{and} \quad BC/DF \to BC/DE.$$

It follows, by the fundamental limit theorem, that

$$\triangle ABC : \triangle ADE = BC : DE,$$

and the incommensurable case is established.

The approach above uses the fact that any irrational number may be regarded as the limit of a sequence of rational numbers, an approach that was rigorously developed in modern times by Georg Cantor (1845–1918).

There is much to recommend the inclusion, in high school demonstrative geometry, of the simple limit procedure illustrated above. For, first of all, the limit concept requires time for assimilation, and the sooner and oftener a student contemplates the concept in significant situations, the better. Next, the theory of limits is essential in any geometric study of areas and volumes. Finally, the inclusion of a discussion of limits in the high school can later benefit the college mathematics student by preparing him for his encounter with the calculus. Though there are postulational approaches to beginning geometry that circumvent the need to consider commensurable and incommensurable cases, the theory of limits is

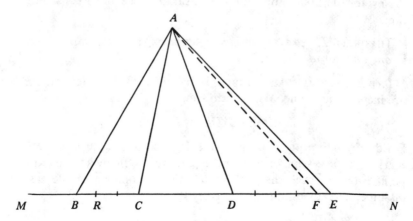

FIG. 23

necessarily required elsewhere in geometry. In particular, as one proceeds in geometry, one discovers that there are many geometrical concepts whose definitions require the limit idea—such as the *length* of a curvilinear arc and the *tangent line* to a curve at a point on the curve.

In a later GREAT MOMENT IN MATHEMATICS we shall see how events of the modern era have, at least so far as mathematics is concerned, vindicated the Pythagorean philosophy that all depends upon the whole numbers. We shall also see that one of the modern methods of completing the real number system is by means of the so-called *Dedekind Schnitt,* or *cut,* which is largely an arithmetized version of the Eudoxian doctrine of ratio and proportion.

Exercises

6.1. Consider the proposition: *Central angles in the same or equal circles are to each other as their intercepted arcs.*

(a) Establish, in Pythagorean fashion, the case where the two central angles are commensurable.

(b) Establish, by the Eudoxian method, both the commensurable and incommensurable cases together.

(c) Establish, with the aid of simple limit notions, the incommensurable case.

6.2.(a) Prove, by the Pythagorean method aided by simple limit notions, that: *A line parallel to one side of a triangle divides the other two sides proportionally.*

(b) Derive the above proposition from the proposition established in the lecture text.

6.3. Prove, by the Pythagorean method aided by simple limit notions, the propositions:

(a) *The areas of two rectangles having equal altitudes are to each other as their bases.*

(b) *Two dihedral angles are to one another as their plane angles.*

(c) *The volumes of two rectangular parallelepipeds having congruent bases are to each other as their altitudes.*

6.4. There are three features of the limit idea that should be noted carefully. If k is the limit of a real variable x, then: (1) k is a constant; (2) the numerical difference between x and k ultimately becomes less than any chosen positive quantity h, however small; (3) the numerical difference between x and k ultimately remains less than the chosen quantity h.

Let A, B, C, three collinear points with B between A and C, represent three railroad stations along a straight track, and let x be the distance from A of a train T traveling toward C. In view of the three features of a limit described above, answer the following questions:

(a) If the train T starts at A and intends to continue until it reaches C, is it proper to say that $\lim x = AB$?

(b) If the train starts at A and intends to run only as far as B, is it proper to say that $\lim x = AC$?

(c) Suppose that a freight train F is traveling from A to C, and that we denote its distance from A by the variable y. Further, suppose the train T leaves A on a parallel track some time after F has pulled out, overtakes F, and then travels alongside F at the same speed as F is traveling. Is it proper to say that $\lim x = y$?

(d) If the train T starts at A and travels at a uniform speed until it reaches the end of the line at C, is it proper to say that $\lim x = AC$?

(e) Suppose the train T starts at A and travels toward C at an ever-decreasing speed, so that in 1 hour it has traveled half the distance from A to C, in the next hour it travels half the remaining distance to C, in the next hour it travels half of what now remains, and so on. Is it proper to say that $\lim x = AC$?

6.5. Establish the *Fundamental limit theorem*.

6.6.(a) Knowing how to measure lengths of straight-line segments, how might one define the length of the circumference of a circle?

(b) Knowing how to measure areas of polygons, how might one define the area of a circle?

6.7. Let P_n and A_n denote, respectively, the perimeter and the area of a regular n-gon circumscribed about a circle of unit diameter. Find $\lim_{n\to\infty} P_n$ and $\lim_{n\to\infty} A_n$.

6.8. Formulate a definition, involving the limit concept, of (a) the tangent line to a curve at a point on the curve, (b) the tangent plane to a surface at a point on the surface.

6.9.(a) Knowing the volume of a pyramid is given by one-third the area of its base times its altitude, how, using limit ideas, might one arrive at a formula for the volume of a circular cone?

(b) Knowing the lateral surface of a regular prism is given by the perimeter of its base times its altitude, how might one define the lateral area of a right circular cylinder?

One might think that one could define the lateral area of a right circular cylinder as the limit of the areas of any sequence of inscribed polyhedral surfaces, provided the number of faces of the polyhedral surfaces indefinitely increases and the maximum face area approaches zero. Mathematicians were surprised when, in the early 1860's, H. A. Schwarz (1843–1921) showed that this is not so. Schwarz's example was so astonishing to the mathematicians of the time that it became known as *Schwarz's paradox.*

6.10. Review, from an elementary calculus text, the geometric interpretation of the derivative of $f(x)$ at x_1 as the slope of the curve $y = f(x)$ at the point (x_1, y_1), where $y_1 = f(x_1)$.

Further Reading

HEATH, T. L., *History of Greek Mathematics.* 2 vols. New York: Oxford University Press, 1931.

ZAMES, FRIEDA, "Surface area and the cylinder area paradox," *Two-Year College Mathematics Journal.* Sept. 1977, pp. 205–211.

FIRST STEPS IN ORGANIZING MATHEMATICS

The Greeks accomplished a great deal in mathematics during the three hundred years following Thales in 600 B.C. Not only did the Pythagoreans and others develop a considerable body of elementary geometry and number theory, but there also evolved notions concerning infinitesimals and summation processes that later, in the seventeenth century, blossomed into the calculus. Also, much higher geometry (that is, the geometry of curves other than the straight line and circle and of surfaces other than the plane and sphere) was developed. Curiously, a great deal of this higher geometry originated in vain attempts to solve three famous challenging problems of antiquity—the duplication of a cube, the trisection of an arbitrary angle, and the quadrature of a circle—illustrating the principle that growth in mathematics is stimulated by the presence of challenging unsolved problems.

But perhaps the greatest accomplishment during the first three hundred years of Greek mathematics was the Greek notion of a logical discourse as a sequence of statements obtained by deductive reasoning from an accepted set of initial statements assumed at the outset of the discourse. Certainly, in the presentation of an argument by deductive procedure, any statement of the argument must be derived from some previous statement or statements of the argument, and such a previous statement must itself be derived from some still more previous statement or statements. Since this cannot be continued backward indefinitely, and since one should not resort to illogical circularity by deriving a statement q from a statement p, and then later deriving statement p from statement q, one must set down at the start of the discourse a collection of primary statements whose truths are to be accepted by the reader, and then to proceed by purely deductive reasoning to derive all the other statements of the discourse.

Now both the primary and the derived statements of the discourse are statements about the technical matter of the discourse, and hence involve special or technical terms. These terms need to be defined. Since technical terms must be defined by means of other technical terms, and these other technical terms by means of still others, one is faced with a difficulty similar to that encountered with the statements of the discourse. In order to get started, and to avoid circularity of definition where term y is defined by means of term x, and then later term x by means of term y, one is again forced to set down, at the very start of the discourse, a collection of basic technical terms, with explanations, the Greeks felt, of their intended usage. All subsequent technical terms of the discourse are then to be defined carefully by means of technical terms already introduced.

A logical discourse, then, according to the Greeks, should be developed according to the following pattern:

Pattern of Material Axiomatics

(A) Initial explanations of certain basic technical terms of the discourse are given, the intention being to suggest to the reader what is to be meant by these basic terms.

(B) Certain primary statements concerning the basic terms, and which are felt to be acceptable to the reader as true on the basis of the properties suggested by the initial explanations, are listed. These primary statements are called the *axioms*, or the *postulates*, of the discourse.

(C) All other technical terms of the discourse are defined by means of previously introduced terms.

(D) All other statements of the discourse are logically deduced from previously accepted or established statements. These derived statements are called the *theorems* of the discourse.

A discourse carried out according to the plan above is today said to be developed by *material axiomatics*. Certainly the most outstanding contribution of the early Greeks to mathematics was the formulation of the pattern of material axiomatics and the insistence that mathematics be systematized according to this pattern. The concept of axiomatic development in mathematics must be ranked as one of the very greatest of the GREAT MOMENTS IN MATHEMATICS.

To acquire a feeling for the pattern of material axiomatics, let us consider an example. Though our example may seem simple and somewhat artificial, nevertheless it is conceivable that it could interest someone, and it will serve our purpose.

The basic, or primitive, terms of our discourse will be a certain (finite and nonempty) collection S of persons and certain clubs formed among these persons. Following the Greek concept of axiomatics, we commence by explaning to the reader just what is to be meant by these primitive terms. By *person* will be meant any man, woman, or child in the collection S, and by *club* will be meant a (nonempty) subset of these persons formed, perhaps, for some civic or other purpose. About these persons and their clubs we shall assume the following:

P1. *Every person of S is a member of at least one club.*

P2. *For every pair of persons of S there is one and only one club to which both belong.*

DEFINITION. Two clubs having no members in common are called *conjugate clubs.*

P3. *For every club there is one and only one conjugate club.*

The three italicized statements above about our primitive terms are the *axioms,* or *postulates,* of the discourse. In view of the preliminary explanations of the primitive terms, the reader should find no difficulty in accepting, for purposes of development, any of these postulates. Note that, for convenience in stating postulate P3 tersely, that postulate was preceded by a definition.

It is possible, by pure deduction, to derive a number of consequences of the above set of postulates. We shall content ourselves with the derivation of four such consequences.

T1. *Every person of S is a member of at least two clubs.*

Let a be a member of S. By P1 there exists a club A to which a belongs. By P3 there exists a club B conjugate to club A. Since B is nonempty, it has at least one member b, and $b \neq a$. By P2 there exists a club C containing a and b. Since A and B are conjugate clubs, b is not in A, implying $A \neq C$. Thus a belongs to two distinct clubs, A and C.

T2. *Every club contains at least two members.*

Let A be a club. Since A is nonempty, it has at least one member a. Suppose a is the only member of A. By T1, there exists a club B distinct from club A and containing a as a member. Now B must contain a second member $b \neq a$, for otherwise clubs A and B would not be distinct. By P3 there exists a club C such that B is conjugate to C. It follows that A is also conjugate to C. But this contradicts P3. Hence the theorem follows by *reductio ad absurdum*.

T3. *S contains at least four persons.*

In the proof of T1 we showed the existence in S of at least two distinct persons a and b. By P2 there exists a club A to which a and b belong. By P3 there exists a club B conjugate to club A. But, by T2, B must contain at least two members, c and d. Since A and B are conjugate, it follows that a, b, c, d are four distinct persons of S.

T4. *There exist at least six clubs.*

In the proof of T3 we showed that there exist four distinct persons a, b, c, d, where c and d belong to a club B conjugate to the club A containing a and b. It now follows that clubs C (containing a and c), D (containing b and d), E (containing a and d), and F (containing c and b) are each distinct from A and B and from each other. Thus there are at least six clubs.

The persevering student may care to try to establish the following much more difficult theorem.

T5. *No club contains more than two members.*

These derived consequences of our postulates are the *theorems* of our discourse. The list can, of course, be considerably extended.

The theory of a simple game can often be developed by material axiomatics. Consider, for example, the familiar game of tic-tac-toe. Among the technical terms of this game are the playing board, nought, cross, a win, a draw, etc. These technical terms are to be explained or defined. The rules of the game are then stated as the postulates of the discourse, these rules being perfectly acceptable once one understands the basic terms of the discourse. From these rules one can then proceed to deduce theorems of the game, proving, for example, that *with proper strategy the player who starts the game need not lose the game.*

There are two theories entertained by historians as to the origin of the axiomatic method—an *evolutionary* one and a *revolutionary* one. The evolutionary theory is the traditional one told by writers living several centuries after the event and seems to be the theory more commonly held today. It conjectures that the axiomatic method gradually evolved as a natural outgrowth and refinement of the early application of deductive procedures in mathematics. The little deductive chains grew longer. Several deductive chains were hooked together in tandem, giving a longer chain yielding intermediate results along the way. And then someone conceived the absolutely grand idea of deducing all of geometry in one single long chain, link by link, starting from some platform of assumption laid down at the start. If this evolutionary theory should be correct, then much of the credit for the axiomatic method must be conferred on the Pythagoreans.

There are some historians of ancient mathematics who find it difficult to believe the evolutionary theory above and who regard the traditional stories as largely legendary. Such an essential turn in the development of mathematics, these historians feel, must have been precipitated by some crucial circumstance that sharply pointed up the importance and delicacy of initial assumption and that led to a whole new level of sophistication. If this was the case, one does not have to search hard for the crucial circumstance—it would have been the devastating crisis in the foundations of mathematics brought on by the discovery of irrational numbers and incommensurable magnitudes. If this revolutionary theory of the origin of the axiomatic method should be the true one, then the credit for the method should, in all likelihood, go to Eudoxus, the genius of the time who resolved the underlying crisis.

Perhaps it is needless to hypothesize about the origin of the axiomatic method. In any event, by the turn of the fourth to the third century B.C., the stage was set for Euclid's magnificent and epoch-making application of the method.

We close by commenting that, in the twentieth century, the material axiomatics of the Greeks underwent a sharpening into what is now called *formal axiomatics*. This is a GREAT MOMENT IN MATHEMATICS of its own, and will be detailed in a later lecture. It will be seen that there are some very significant differences between the older and the newer viewpoints of the axiomatic method.

Exercises

7.1. The definition of a technical term (beyond the primitive ones) of a logical discourse serves merely as an abbreviation for a complex combination of terms already present. Thus a new term introduced by definition is really arbitrary, though convenient, and may be entirely dispensed with; but then the discourse in which the vocabulary is to be employed would immediately increase in complexity. Consider, for example, the following definitions taken from an elementary geometry text: (1) The *diagonals* of a quadrilateral are the two straight line segments joining the two pairs of opposite vertices of the quadrilateral. (2) *Parallel lines* are straight lines lying in the same plane and never meeting, however far they are extended in either direction. (3) A *parallelogram* is a quadrilateral having its opposite sides parallel.

Now, without using any of the italicized words above, restate the proposition: "The diagonals of a parallelogram bisect each other."

7.2. In elementary algebra, if n represents a positive integer and if k represents any real number, we define k^n as a symbol to represent the product $(k)(k) \cdots (k)$, in which k appears as a factor n times. Rewrite, without using exponents, the following expression in which a, b, c represent real numbers:

$$[(a + b)^5 (a - b)^3]^7.$$

7.3. Give the customary definitions of the following mathematical symbols and illustrate their convenience:

(a) $\sum_{n=1}^{\infty} a_n$

(b) $n!$, n a positive integer

(c) $\binom{m}{n}$, m and n positive integers with $m \geq n$.

7.4. By means of appropriate definitions reduce the following sentence to one containing not more than five words: "The movable seats with four legs were restored to a sound state by the person who takes care of the building."

7.5. A dictionary must, of necessity, resort to circularity, but it is hoped that any person using the dictionary has developed an adequate vocabulary so that the words in terms of which some unknown word is defined are already familiar to him. Trace the following words through a standard dictionary until a circular chain has been observed: (a) dead, (b) noisy, (c) line (in the mathematical sense).

7.6. Explain how the following outdated, but still commonly heard, statement reflects part of the pattern of Greek material axiomatics: "An axiom is a self-evident truth."

7.7. Using the same basic terms as in the example of material axiomatics developed in the lecture, let us assume:

P1. *Any two distinct clubs have one and only one member in common.*

P2. *Each person of S belongs to two and only two clubs.*

P3. *There are exactly four clubs.*

From these postulates deduce the following theorems:

T1. *There are exactly six persons in S.*

T2. *There are exactly three persons in each club.*

T3. *For each person in S there is exactly one other person in S not in the same club.*

7.8. Establish the theorem, cited in the lecture text, about the game of tic-tac-toe.

7.9. The *cigar game* is played by two players on a rectangular table top with a large stock of cigars. The cigars are assumed to be symmetrical and to be all alike. The two players, taking turns, lay (at each turn) a cigar on the table top so that it does not overlap any other cigar or protrude over the edge of the table top. The last player able to place a cigar on the table wins the game. Prove the following theorem about this game: *Using a suitable strategy, the player who starts the game can win the game.*

7.10. *A, B,* and *C* are playing two simultaneous games of chess, *A* against *B* and *B* against *C*. *A* and *C* are experts and *B* is a mere novice. *A* has white in the one game and *B* has white in the other game. Show that using a clever strategy *B* can either win one game while losing the other, or draw both games.

Further Reading

WILDER, R. L., *Introduction to the Foundations of Mathematics,* 2nd ed. New York: John Wiley, 1965.

THE MATHEMATICIANS' BIBLE

Following the death of Alexander the Great in 323 B.C., the vast Macedonian Empire was divided into three parts, and the part containing Egypt came under the able governance of Alexander's talented general, Ptolemy Soter, who shortly assumed kingship of the region. Ptolemy chose Alexandria, only a few miles removed from the mouth of the Nile River, as his capital, and about 300 B.C. he opened the doors of the famous University of Alexandria. In the galaxy of scholars invited to staff the new institution was the mathematician Euclid, probably a one-time attendant at the Platonic Academy in Athens.

One of the first mathematical tasks undertaken by Euclid upon assumption of his duties at Alexandria was that of assembling his monumental and historically important *Elements*. This very remarkable and extensive work, written in thirteen books, or parts, is the earliest application of the axiomatic method that has come down to us. It is generally regarded as the first great landmark in the history of mathematical organization, and its subsequent influence on scientific writing can hardly be overstated.

Euclid's *Elements* is an outstanding candidate as a GREAT MOMENT IN MATHEMATICS. The work so quickly and so completely superseded all previous works of the same nature that now no copies of the earlier efforts remain and we merely know of their prior existence through commentaries by later writers.* From its first appearance, Euclid's *Elements* was accorded the highest respect. With the unique exception of the Bible, no work has been

---

*David Hilbert once remarked that one can judge the significance of a work by the number of earlier works rendered superfluous.

more widely used, studied, or edited, and for over two millennia it has dominated all teaching of geometry. More than a thousand editions of the work have appeared since the first printed one in 1482. Its content and its form have made a tremendous impact on the development of both the subject matter and the logical foundations of mathematics.

Proclus, a later commentator of mathematics living in the fifth century, has clarified for us the meaning of the term "elements." The elements of a demonstrative study are the leading, or key, theorems which are of frequent and wide use in the subject; they are the theorems required for the proofs of all or most other theorems. Their function has been compared to that of the letters of the alphabet in relation to language; indeed, letters are called by the same name in Greek. The choice of the propositions to serve as the elements of a subject demands considerable skill and judgment on the part of the writer.

It is no discredit to the brilliance of Euclid's achievement that efforts anterior to his had been made. According to Proclus, the first attempt at an *Elements* was made by Hippocrates of Chios in the middle of the fifth century B.C. The next effort was that of Leon, coming in age somewhere between Plato and Eudoxus. Leon's work, we are told, contained a larger and more serviceable selection of propositions than did the work of Hippocrates. The textbook employed at Plato's Academy was assembled by Theudias of Magnesia and was regarded as an admirable collection of elements. The work of Theudias was apparently the immediate precursor of Euclid's work and was undoubtedly available to Euclid, especially if, as is believed, he studied in the Platonic Academy. Euclid was acquainted also with the work of Theatetus and Eudoxus. But it is no detraction that Euclid's work is, to a great extent, a compilation of works of his predecessors, for the chief merit in Euclid's *Elements* lies precisely in the consummate skill with which the propositions were selected and arranged in a logical sequence, presumably following from a small handful of initial assumptions. Nor is it a detraction that the searchlight of modern criticism has revealed certain defects in the structure of Euclid's work; one could hardly expect such an early and colossal attempt by the axiomatic method to be free of all blemishes.

There is no extant copy of Euclid's *Elements* dating from the author's own time. All modern editions of the *Elements* are based upon a revision that was prepared by Theon of Alexandria, a Greek commentator who lived almost seven hundred years after the time of Euclid. It was not until the beginning of the nineteenth century that an older copy was discovered. In 1808, when Napoleon ordered valuable manuscripts to be taken from Italian libraries and to be sent to Paris, F. Peyrard found, in the Vatican library, a tenth-century copy of an edition of Euclid's *Elements* that predates the Theon recension. But a study of this older edition has shown only minor differences from Theon's edition.

The first complete translations of the *Elements* into Latin were not made from the Greek but from the Arabic. In the eighth century a number of Byzantine manuscripts of Greek works were translated by the Arabians; and in 1120 the English scholar, Adelard of Bath, made a Latin translation of the *Elements* from one of these older Arabian translations. Other Latin translations from the Arabic were made by Gherardo of Cremona (1114–1187) and, 150 years after Adelard, by Johannes Campanus. The first printed edition of the *Elements* was made at Venice in 1482 and contained the Campanus translation. This very rare book was beautifully executed and was the first mathematical book of significance to be printed. An important Latin translation from the Greek was made by Commandino in 1572. This translation served as a basis for many subsequent translations, including the very influential work by Robert Simson, from which, in turn, so many of the English editions were derived.

The famous French mathematician, Adrien-Marie Legendre (1752–1833), known in the history of mathematics chiefly for his work in number theory, elliptic functions, the method of least squares, and integrals, was also interested in pedagogical matters. In his very popular *Éléments de géométrie,* he attempted a pedagogical improvement of Euclid's *Elements* by considerably rearranging and simplifying many of Euclid's propositions. This work was very favorably received in America and became the prototype of the geometry books in this country. In fact, the first English translation of Legendre's geometry was made in 1819 by

John Farrar of Harvard University. Three years later another English translation was made by the famous Scottish littérateur, Thomas Carlyle, who early in life was a teacher of mathematics. Carlyle's translation, as later revised by others, ran through 33 American editions. Until quite recent times, most American high school geometry textbooks were fashioned after the Legendre revision.

In the Theon edition, Euclid's *Elements* comprises thirteen books, or parts, and contains a total of 465 propositions. Contrary to popular impression, much of the material is concerned, not with geometry, but with elementary number theory and Greek algebra.

Book I commences with the necessary preliminary definitions and explanations, postulates, and axioms. Though today mathematicians use the words "axiom" and "postulate" synonymously, some of the early Greeks made a distinction, the distinction adopted by Euclid being, it seems, that an axiom is an initial assumption common to all fields of study, whereas a postulate is an initial assumption peculiar to the particular study at hand. Among the propositions of Book I are the familiar theorems on congruence, parallel lines, and rectilinear figures. Propositions 47 and 48, the two final propositions of the book, are the Pythagorean theorem and its converse, and one is reminded of a story told of the English philosopher Thomas Hobbes (1588–1679). One day, opening Euclid's *Elements*, by chance, at the theorem of Pythagoras, Hobbes exclaimed, "By God, this is impossible," and then proceeded to read the proofs of Book I in retrograde order until, arriving at the axioms and postulates, he became convinced.

Book II, a short book of only 14 propositions, deals largely with the geometric algebra of the Pythagorean school. We have already noted, in LECTURE 4, that Propositions 12 and 13 of this book are essentially the generalization of the Pythagorean theorem known today as the *law of cosines.*

Book III, consisting of 39 propositions, contains the familiar theorems about circles, chords, secants, tangents, and the measurement of associated angles that we find in our current high school geometry books. Book IV, with only 16 propositions, is devoted to the construction, with straightedge and compasses, of certain regular polygons, their inscription within a given circle, and their circumscription about a given circle.

Book V, as we remarked in the preceding lecture, is devoted to a masterly exposition of the Eudoxian theory of proportion. This book is regarded as one of the greatest masterpieces of mathematical literature. It was the reading of this book, it will be recalled from our previous lecture, that cured Bolzano of the physical distress he once suffered while vacationing in Prague. Book VI, one of the richest books of the *Elements,* applies the Eudoxian theory to the study of similar figures.

Books VII, VIII, and IX, containing a total of 102 propositions, deal with elementary number theory. Book VII commences with the process, referred to today as the *Euclidean algorithm,* for finding the greatest common integral divisor of two or more integers. We also find an exposition of the early Pythagorean theory of proportion. Book VIII concerns itself largely with continued proportions and related geometric progressions. If we have the continued proportion $a:b = b:c = c:d,$ then a, b, c, d form a geometric progression. There are many significant theorems in number theory in Book IX. Proposition IX 14 is equivalent to the important *fundamental theorem of arithmetic,* namely, that *any integer greater than* 1 *can be expressed as a product of primes in one, and essentially only one, way.* In Proposition IX 20 we find a singularly elegant proof of the fact that *the number of prime numbers is infinite.* Proposition IX 35 furnishes a geometric derivation of the formula for the sum of the first n terms of a geometric progression, and the last proposition of the book, IX 36, establishes a remarkable formula yielding even perfect numbers.

Book X, a difficult book to read, concerns itself with irrationals—that is, with line segments which are incommensurable with respect to some given line segment. The remaining three books, XI, XII, and XIII, concern themselves with solid geometry, covering much of the material, with the exception of that on spheres, commonly found in present-day high school texts. We shall see, in our next lecture, that the fundamental material on spheres was supplied, some time later, by Archimedes.

The material found in current American high school plane and solid geometry texts is largely that found in Euclid's Books I, III, IV, VI, XI, and XII. The material in the current high school texts concerning the measurement of the circle and the sphere, and the material in solid geometry dealing with spherical triangles, is of later origin and is not found in Euclid's *Elements*.

There are other great treatises of Greek antiquity, besides Euclid's *Elements,* that have come down to us. Thus there are the profound works of Archimedes, Apollonius' *Conic Sections,* Ptolemy's *Almagest,* Heron's *Metrica,* Diophantus' *Arithmetica,* Menelaus' *Sphaerica,* Pappus' *Mathematical Collection,* and others, and, in a more extended series of lectures, all or many of these might merit inclusion in a selection of GREAT MOMENTS IN MATHEMATICS. In our present curtailed selection, we shall limit ourselves to the consideration of only a few of these remarkable works. But, no matter how brief a selection of GREAT MOMENTS IN MATHEMATICS might be made, Euclid's *Elements* would surely be present.

Because of important subsequent consequences, which we shall consider in later lectures, we here close our present lecture by listing Euclid's axioms and postulates:

The Postulates

1. A straight line segment may be drawn connecting any two given points.
2. A straight line segment may be produced continuously in a straight line in either direction.
3. A circle may be drawn with any given point as center and passing through any given second point.
4. All right angles are equal to one another.
5. If a straight line falling on two straight lines makes the interior angles on the same side together less than two right angles, the two straight lines, if produced indefinitely, meet on that side on which the angles are together less than two right angles.

The Axioms or Common Notions

1. Things which are equal to the same thing are equal to one another.

2. If equals be added to equals, the sums are equal.

3. If equals be subtracted from equals, the remainders are equal.

4. Things which coincide with one another are equal to one another.

5. The whole is greater than the part.

We note that the first three postulates restrict constructions to those that can be made with compasses and an unmarked straight-edge. It is for this reason that these two instruments are often referred to as the *Euclidean tools,* and constructions performable by them as *Euclidean constructions.* Euclid used constructions in the sense of existence theorems, to prove that certain entities actually exist. Thus one may define a *bisector* of a given angle as a line in the plane of the angle, passing through the vertex of the angle, and such that it divides the given angle into two equal angles. But a definition does not guarantee the existence of the thing defined; this requires proof. To show that a given angle does possess a bi-sector, we may show that this entity can actually be constructed. Existence theorems are very important in mathematics, and actual construction of an entity is the most satisfying way of proving its existence. One might define a *square circle* as a figure which is both a square and a circle, but one would never be able to prove that such an entity exists; the class of square circles is a class without any members. In mathematics, it is nice to know that the set of entities satisfying a certain definition is not just the empty set.

The fact that there are constructions beyond the Euclidean tools has added zest to this facet of geometry, and vain attempts to ob-tain constructions now known to be impossible with these limited tools led to the discovery of a considerable amount of interesting geometry.

We shall later see that Euclid's fifth postulate led, in the nine-teenth century, to consequences of tremendous importance for the development of mathematics. Cassius J. Keyser has dubbed the postulate "the most famous single utterance in the history of sci-ence."

Exercises

8.1. If you are to choose two of the following theorems for "elements" of a course in plane geometry, which would you choose?

1. The three altitudes of a triangle, produced if necessary, meet in a point.
2. The sum of the three angles of a triangle is equal to two right angles.
3. An angle inscribed in a circle is measured by half its intercepted arc.
4. The tangents drawn from any point on the common chord produced of two given intersecting circles are equal in length.

8.2.(a) A geometry teacher is going to present the topic of parallelograms to her class. After defining *parallelogram*, what theorems about parallelograms should the teacher offer as the "elements" of the subject?

(b) Preparatory to teaching the topic on similar figures, a geometry teacher gives a lesson or two on the theory of proportion. What theorems should she select for the "elements" of the treatment, and in what order should she arrange them?

8.3.(a) A mathematics instructor is going to present the subject of geometric progressions to his college algebra class. After defining this type of progression, what theorems about geometric progressions should the instructor offer as the "elements" of the subject?

(b) Imagine yourself building up an elementary treatment of trigonometric identities. Which identities would you select for the "elements" of your treatment, and in what order would you arrange them?

8.4. As an illustration of nongeometrical material found in Euclid's *Elements*, let us consider the *Euclidean algorithm*, or process, for finding the greatest common integral divisor (g.c.d.) of two positive integers. The process is found at the start of Euclid's Book VII, although it was no doubt known before Euclid's time.

The algorithm lies at the foundation of several developments in modern mathematics. Stated in the form of a rule the process is this: *Divide the larger of the two positive integers by the smaller one. Then divide the divisor by the remainder. Continue this process, of dividing the last divisor by the last remainder, until the division is exact. The final divisor is the sought g.c.d. of the two original positive integers.*

(a) Find, by the Euclidean algorithm, the g.c.d. of 5913 and 7592.

(b) Find, by the Euclidean algorithm, the g.c.d. of 1827, 2523, and 3248.

(c) Prove that the Euclidean algorithm does lead to the g.c.d.

(d) Let h be the g.c.d. of the positive integers a and b. Show that there exist integers p and q (not necessarily positive) such that $pa + qb = h$.

(e) Find p and q for the integers of part (a).

(f) Prove that a and b are relatively prime if and only if there exist integers p and q such that $pa + qb = 1$.

8.5.(a) Prove, using Exercise 8.4 (f), that if p is a prime and divides the product uv, then either p divides u or p divides v.

(b) Prove, using part (a), the *fundamental theorem of arithmetic: Every integer greater than 1 can be uniquely factored into a product of primes.*

(c) Find integers a, b, c such that $65/273 = a/3 + b/7 + c/13$.

8.6. The fundamental theorem of arithmetic says that, for any given positive integer a, there are unique nonnegative integers $a_1, a_2, a_3, \ldots$, only a finite number of which are different from zero, such that

$$a = 2^{a_1} 3^{a_2} 5^{a_3} \ldots,$$

where $2, 3, 5, \ldots$ are the consecutive primes. This suggests a useful notation. We shall write

$$a = (a_1, a_2, \ldots , a_n),$$

where a_n is the last nonzero exponent. Thus we have $12 = (2, 1)$, $14 = (1, 0, 0, 1)$, $27 = (0, 3)$, and $360 = (3, 2, 1)$.

Prove the following theorems:

(a) $ab = (a_1 + b_1, a_2 + b_2, \dots)$.

(b) b is a divisor of a if and only if $b_i \le a_i$, for each i.

(c) The number of divisors of a is $(a_1 + 1)(a_2 + 1) \cdots (a_n + 1)$.

(d) A necessary and sufficient condition for a number a to be a perfect square is that the number of divisors of a be odd.

(e) Set g_i equal to the smaller of a_i and b_i if $a_i \ne b_i$ and equal to either a_i or b_i if $a_i = b_i$. Then $g = (g_1, g_2, \dots)$ is the g.c.d. of a and b.

(f) If a and b are relatively prime and b divides ac, then b divides c.

(g) If a and b are relatively prime and if a divides c and b divides c, then ab divides c.

(h) Show that $\sqrt{2}$ and $\sqrt{3}$ are irrational.

8.7. Euclid defines a circle as "a plane figure contained by one line such that all the straight lines falling upon it from one particular point among those lying within the figure are equal." How does this definition of a circle differ from the modern definition of a circle?

8.8. One should understand precisely the intention of Euclid's Postulate 3. It says that, given two points A and B, one can draw the circle with center A and radius AB. It follows that the Euclidean compasses differ from our modern compasses, for with the modern compasses one is permitted to draw a circle having any point A as center and any segment BC as radius. In other words, one is permitted to transfer the distance BC to the center A, using the compasses as dividers. The Euclidean compasses, on the other hand, may be supposed to collapse if either leg is lifted from the paper.

A student reading Euclid's *Elements* for the first time might experience surprise at the opening propositions of Book I. The first three propositions are the construction problems:

I 1. *To describe an equilateral triangle upon a given line segment.*

I 2. *From a given point to draw a line segment equal to a given line segment.*

I 3. *From the greater of two given line segments to cut off a part equal to the lesser.*

These three constructions are trivial with straightedge and *modern* compasses, but require some ingenuity with straightedge and *Euclidean* compasses.

(a) Solve construction I 1 with Euclidean tools.

(b) Solve construction I 2 with Euclidean tools.

(c) Solve construction I 3 with Euclidean tools.

(d) Show that Proposition I 2 proves that the straightedge and *Euclidean* compasses are equivalent to the straightedge and *modern* compasses.

8.9. In Book II of Euclid's *Elements* a number of algebraic identities are established in a geometric fashion. Show how each of the following identities might be so established, assuming a, b, c, d are positive quantities:

(a) $(a + b)^2 = a^2 + 2ab + b^2$

(b) $(a - b)^2 = a^2 - 2ab + b^2, a > b$

(c) $a^2 - b^2 = (a + b)(a - b), a > b$

(d) $a(b + c) = ab + ac$

(e) $(a + b)^2 = (a - b)^2 + 4ab, a > b$

(f) $(a + b)(c + d) = ac + bc + ad + bd$.

8.10.(a) Let r and s denote the roots of the quadratic equation

$$x^2 - px + q^2 = 0,$$

where p and q are positive numbers. Show that $r + s = p$, $rs = q^2$, and r and s are both positive if $q \le p/2$.

(b) To solve the quadratic equation of part (a) geometrically for real roots, we must find line segments r and s from given line segments p and q. That is, we must construct a rectangle equivalent to a given square and having the sum of its base and altitude equal to a given line segment. Devise a suitable construction based on Figure 24, and show geometrically that for real roots to exist we must have $q \le p/2$.

(c) Let r and s denote the roots of the quadratic equation

$$x^2 - px - q^2 = 0,$$

where p and q are positive numbers. Show that $r + s = p$, $rs = -q^2$, the roots are real, and the numerically larger one is positive while the other is negative.

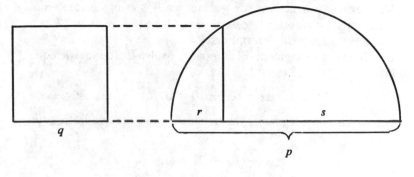

FIG. 24

(d) To solve the quadratic equation of part (c) geometrically, we must find line segments r and s from given line segments p and q. That is, we must construct a rectangle equivalent to a given square and having the difference of its base and altitude equal to a given line segment. Devise a suitable construction based on Figure 25.

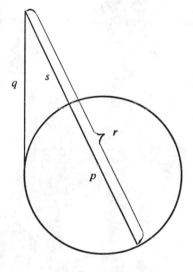

FIG. 25

(e) Devise constructions for geometrically solving for real roots the quadratic equations $x^2 + px + q^2 = 0$ and $x^2 + px - q^2 = 0$, where p and q are positive numbers.

(f) Given a unit segment, geometrically solve the quadratic equation

$$x^2 - 7x + 10 = 0.$$

(g) Given a unit segment, geometrically solve the quadratic equation

$$x^2 - 4x - 21 = 0.$$

(h) With straightedge and compasses, divide a segment a into two parts such that the difference of their squares shall be equal to their product.

(i) Show that, in part (h), the longer segment is a mean proportional between the shorter segment and the whole segment. (The line segment is said to be divided in *extreme and mean ratio*, or in *golden section*. See LECTURE 5.)

Further Reading

HEATH, T. L., *The Thirteen Books of Euclid's Elements*, 2nd ed., 3 vols. New York: Cambridge University Press, 1926. Reprinted by Dover, 1956.

LECTURE **9**

THE THINKER AND THE THUG

It is surprising how many of the great developments of mathematics of modern times find their origins in work done two millennia earlier by the ancient Greeks. As Julian Lowell Coolidge, the eminent Harvard geometer of the early twentieth century, was fond of saying: "There were giants in the land then." Beyond any doubt the very greatest of all those giants was the famous Archimedes of Syracuse, a man of incredible mathematical talent. In this lecture we shall see that Archimedes, in finding, with the limited means at his disposal, areas of certain curvilinear plane figures and areas and volumes of certain curved surfaces, gave birth to the integral calculus that many, many years later was to evolve toward perfection under Kepler, Cavalieri, Fermat, Wallis, Barrow, Leibniz, and Newton.

Archimedes, a native of the Greek city of Syracuse on the island of Sicily, was born about 287 B.C. and died during the Roman pillage of Syracuse in 212 B.C. He was the son of an astronomer and in high favor with King Hieron of Syracuse. It is believed that he spent some years at the University of Alexandria in Egypt, for he numbered Conon, Dositheus, and Eratosthenes among his friends, all three members of the famous institution. Many of Archimedes' mathematical discoveries were communicated to these able successors of Euclid.

Many colorful stories about Archimedes have been told by later Roman historians. Thus there are descriptions of the cunning contrivances that Archimedes devised to assist in the defense of Syracuse during the two-year siege directed by the frustrated Roman general Marcellus. There were catapults with adjustable ranges, projecting poles on bases that could be quickly wheeled to any part

83

of the city's walls and there drop heavy weights on approaching enemy ships, and huge movable grappling cranes that could lift any invading vessel from the water and shake it into pieces. The story that Archimedes employed large burning-glasses to set the sails of outlying enemy ships afire is of later origin, but could be true. There is also the story of how Archimedes lent credence to his boast, "Give me a place to stand on and I will move the earth," by effortlessly and single-handedly, while comfortably seated in a chair on the beach, moving with a compound pulley a heavily weighted ship into the water from dry dock.

Like some other great mathematicians, Archimedes was capable of strong mental concentration, and stories are told of his obliviousness to surroundings when engrossed by a problem. Often told is the story about King Hieron and a suspected goldsmith. King Hieron had a goldsmith fashion him a gold crown, but when the task was finished the king feared the goldsmith had replaced some of the gold by hidden silver. Not wanting to break the crown apart to settle the matter, the king referred the problem to Archimedes, who, one day while in the city's baths, hit upon the key to a solution by discovering the first law of hydrostatics: *A body immersed in a fluid is buoyed up by a force equal to the weight of the displaced fluid.* In his excitement of discovery, he rose from his bath and, forgetting to clothe himself, ran home through the streets shouting, "Eureka, eureka."*

Archimedes made many of his geometrical discoveries from figures drawn in the ashes of his fireplace or in the coating of after-bathing oil that he smeared on his body. In fact, Roman historians have related that he met his end while preoccupied with a geometrical diagram drawn on a sand tray when, during a careless relaxation of the watches, Marcellus and his troops finally broke into the besieged city. According to one version of the story, when the

---

*An event now amusingly preserved for us in a clerihew, given by J. C. W. de La Bere in the December 1974 issue of the *Australian Mathematical Society Gazette:*

> Archimedes of Syracuse
> To get into the news,
> Called out "Eureka"
> And became the first streaker.

shadow of a pillaging Roman soldier fell across his diagram, Archimedes waved the intruder back with a command not to disturb his figure, whereupon the incensed looter ran a spear through the old man.

Ten treatises of Archimedes have come down to us, and there are traces of some lost works. The extant treatises are all masterpieces of mathematical exposition, written with a high finish and economy of presentation, and exhibiting great originality, computational skill, and rigor in demonstration. Perhaps the most notable contribution to mathematics made in these works is the early development of the processes of the integral calculus. We now turn our attention to this remarkable attainment.

Several of Archimedes' papers employ procedures that are equivalent to the performance of a genuine integration. We shall here concern ourselves with only two of these papers—Archimedes' own favorite work, *On the Sphere and Cylinder,* and an only recently found treatise entitled *Method.* In the first of these two works, written in two books and containing a total of 60 propositions, appear, for the first time in mathematics, correct expressions for the areas of a sphere and a spherical zone of one base and the volumes of a sphere and a spherical segment of one base. The situations for the area and volume of a sphere are dramatically set out in a corollary to Propositions 33 and 34 of Book I: *The cylinder whose base is equal to a great circle of the sphere and whose altitude is equal to a diameter of the sphere has a total surface* (the lateral surface plus the two bases) *exactly equal to 3/2 the surface of the sphere and a volume exactly equal to 3/2 the volume of the sphere.* From this it is easy to obtain the familiar formulas

$$S = 4\pi r^2 \quad \text{and} \quad V = 4\pi r^3/3$$

for the surface area S and volume V of a sphere of radius r. It is in the chain of propositions ingeniously leading step by step to these results that integration ideas are involved, where, in place of the direct and supple modern method of limits, we find the indirect and cumbersome, but equally serviceable, double *reductio ad absurdum* procedure known as the *Eudoxian method of exhaustion.* Although, in a more extended sequence of lectures, the Eudoxian method of exhaustion would be included as a GREAT MOMENT IN MATHE-

MATICS, we here forego an exposition of that method and, therefore, along with it, a proper examination of Archimedes' clever establishment of the two formulas above for the area and volume of a sphere.* Instead, we shall examine in some detail an account given by Archimedes in his work *Method*, wherein he tells us how he induced the concerned formulas in the first place. This account not only clearly involves integration ideas, but also supplies an interesting method of discovery.

Archimedes' *Method* had been a long-lost treatise, known only by references to it, until the discovery of a tenth-century copy of it in 1906 in Constantinople by the distinguished German historian of mathematics J. L. Heiberg. The found work was a palimpsest—that is, a parchment which some centuries later had been reused for other writing by washing off the original ink, but on which, with the passage of time, the first writing dimly reappeared beneath the later writing. The work is in the form of a letter addressed to Eratosthenes at the University of Alexandria.

The method of exhaustion, though rigorous, is a sterile method. That is, once a formula is known, the method of exhaustion may furnish an elegant tool for establishing the formula, but the method does not lend itself to the initial discovery of the result.† How, then, did Archimedes discover, for example, the formulas found in his treatise *On the Sphere and Cylinder* that he so neatly established there by the method of exhaustion? The answer to this question is given by Archimedes in his work *Method*.

The fundamental idea of Archimedes' method (known more fully as his *method of equilibrium*) is this: To find a required area or volume, cut it up into a very large number of thin parallel planar strips, or thin parallel slices, and (mentally) hang these pieces at one end of a given lever in such a way as to be in equilibrium with a figure whose content and centroid are known. Let us illustrate the

---

*We may point out, however, that essentially Archimedes' treatment, refined by the modern theory of limits, can be found in almost any high school text on solid geometry.

†In this respect, the method of exhaustion is very much like the process of *mathematical induction* that a student encounters in his algebra course in high school.

method by using it to discover the formula for the volume of a sphere.

Let r be the radius of the sphere. Place the sphere with its polar diameter along a horizontal x-axis with the north pole N at the origin (see Figure 26). Construct the cylinder and the cone of revolution obtained by rotating the $2r \times r$ rectangle *NABS* and the triangle *NCS* (which is a right isosceles triangle with legs of length $2r$) about the x-axis. Now cut from the three solids thin vertical slices (assuming that they are flat cylinders) at distance x from N and of thickness Δx. The volumes of these slices are, approximately,

$$
\begin{aligned}
\text{sphere:} \quad & \pi x(2r - x)\Delta x,* \\
\text{cylinder:} \quad & \pi r^2 \Delta x, \\
\text{cone:} \quad & \pi x^2 \Delta x.
\end{aligned}
$$

Take corresponding slices from the sphere and cone and hang them with their centers at T, where $TN = 2r$. The combined moment † of these two slices about N is

$$[\pi x(2r - x)\Delta x + \pi x^2 \Delta x]2r = 4\pi r^2 x \Delta x.$$

This, we observe, is four times the moment of the slice cut from the cylinder when that slice is left where it is. Adding a large number of these slices together we find

$2r$[volume of sphere + volume of cone] = $4r$[volume of cylinder],

or

$$2r\left[\text{volume of sphere} + \frac{8\pi r^3}{3}\right] = 8\pi r^4,$$

or

$$\text{volume of sphere} = \frac{4\pi r^3}{3}.$$

This, we are told in *Method*, was Archimedes' way of discovering the formula for the volume of a sphere. But his mathematical conscience

---

*Since the radius of this slice is the mean proportional between x and $2r - x$.

†The *moment* of a volume about a point is the product of the volume and the distance from the point to the centroid of the volume.

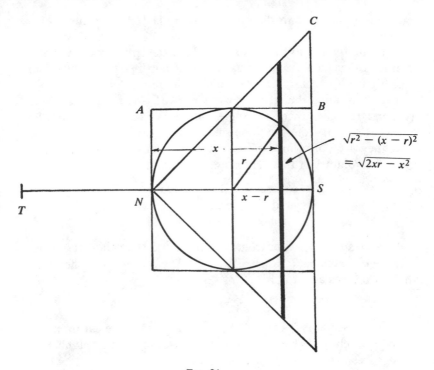

FIG. 26

would not permit him to accept such a method as a proof, and he accordingly supplied a rigorous demonstration by means of the method of exhaustion.

In Archimedes' method of equilibrium we see the fertility of the loosely founded idea of regarding a magnitude as composed of a large number of atomic pieces. Needless to say, with the modern theory of limits, Archimedes' method of equilibrium can be made perfectly rigorous, and becomes essentially the same as present-day integration. Surely, from almost any point of view, we have here in Archimedes' work a truly GREAT MOMENT IN MATHEMATICS.

Archimedes was so fond of his treatise *On the Sphere and Cylinder* that he said that when he died he would like to have engraved on his tombstone the figure of a sphere inscribed in a cylinder, as in Figure 27. Now Marcellus had built up such an ad-

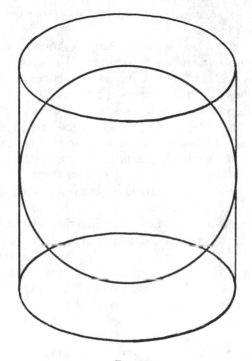

FIG. 27

miration and respect for Archimedes, as his long-successful adversary during the siege of Syracuse, that when he learned Archimedes had perished during the downfall of the city, he had Archimedes buried with great pomp and ceremony and erected over Archimedes' grave a tombstone with the requested figure engraved upon it. Years later, when Cicero, serving as Roman quaestor, visited Syracuse to collect taxes, no one could direct him to Archimedes' grave. After a considerable search, he located the tombstone amidst overgrown brambles and he restored the grounds around the grave site. But, with the passage of time, the grave was again neglected, and seemed, with the growth of the city over the centuries, to be irretrievably lost. However, in 1965, while excavating for the foundations of a new hotel in Syracuse, the steam shovel came up bearing a tombstone with the figure of a sphere inscribed in a cylinder engraved upon it. So once again, the tomb of the greatest of all Syracusans was found.

The Czechoslovakian scientist Petr Beckmann, now of the Department of Electrical Engineering at the University of Colorado, sees history as made largely by a mortal conflict between two classes of people of the world, the thinkers and the thugs. About these two classes Beckmann has fashioned what may be called *Beckmann's law*: "In the conflict between the thinkers and the thugs, the thugs always win, but the thinkers always outlive them." For example, the Greeks, and Archimedes in particular, were thinkers, while the Romans, and Marcellus in particular, were thugs. In the contest between Archimedes and Marcellus, Marcellus won, but the accomplishments of Achimedes will long outlast anything done by Marcellus. It was about this fateful conflict that Sir William Rowan Hamilton, Ireland's greatest bid to glory in the field of mathematics, once remarked: "Who would not rather have the fame of Archimedes than that of his conqueror Marcellus?" Again, recalling the death of Archimedes, the British philosopher Alfred North Whitehead observed: "No Roman ever died in contemplation over a geometrical diagram." And, though Beckmann's law guarantees human memory of Archimedes to outlive human memory of Marcellus, the eminent British number theorist G. H. Hardy assures us that even among the Greek thinkers themselves, "Archimedes will be remembered when Aeschylus is forgotten, because languages die and mathematical ideas do not."

In a more protracted set of lectures on GREAT MOMENTS IN MATHEMATICS, Archimedes could easily figure more than once. It was Archimedes, for instance, who started the long history of the scientific computation of π, and it was Archimedes who wrote the first significant papers in mathematical physics. In a later lecture we shall return to Book II of Archimedes' *On the Sphere and Cylinder*, to make a comment in connection with the solution of cubic equations.

Exercises

9.1. The first law of hydrostatics appears as Proposition 7 of Book I of Archimedes' work *On Floating Bodies*.

(a) Let a crown of weight w pounds be made up of w_1 pounds of gold and w_2 pounds of silver. Suppose that w pounds of pure gold loses f_1 pounds when weighed in water, that w pounds of pure silver loses f_2 pounds when weighed in water, and that the crown loses f pounds when weighed in water. Show that

$$\frac{w_1}{w_2} = \frac{f_2 - f}{f - f_1}.$$

(b) Suppose the crown of part (a) displaces a volume of v cubic inches when immersed in water, and that lumps, of the same weight as the crown, of pure gold and pure silver displace, respectively, v_1 and v_2 cubic inches when immersed in water. Show that

$$\frac{w_1}{w_2} = \frac{v_2 - v}{v - v_1}.$$

9.2. Obtain, from Archimedes' corollary to Propositions 33 and 34 of Book I of his *On the Sphere and Cylinder*, the familiar formulas for the surface area and volume of a sphere of radius r.

9.3. Define *spherical zone* (of one and two bases), *spherical segment* (of one and two bases), and *spherical sector*.

9.4. Assuming the theorem: *The area of a spherical zone is equal to the product of the circumference of a great circle and the altitude of the zone,* obtain the familiar formula for the area of a sphere and establish the theorem: *The area of a zone of one base is equal to that of a circle whose radius is the chord of the generating arc.*

9.5. Assuming that the volume of a spherical sector is given by one-third the product of the area of its base and the radius of the sphere, obtain the following results:

(a) The volume of a spherical segment of one base, cut from a sphere of radius r, having h as altitude and a as the radius of its base, is given by

$$V = \pi h^2 \left(r - \frac{h}{3} \right) = \pi h \left(\frac{3a^2 + h^2}{6} \right).$$

(b) The volume of a spherical segment of two bases, having h as altitude and a and b as the radii of its bases, is given by

$$V = \frac{\pi h(3a^2 + 3b^2 + h^2)}{6} .$$

(c) The spherical segment of part (b) is equivalent to the sum of a sphere of radius $h/2$ and two circular cylinders whose altitudes are each $h/2$ and whose radii are a and b, respectively.

9.6. Figure 28 represents a parabolic segment having AC as chord. CF is tangent to the parabola at C and AF is parallel to the axis of the parabola. OPM is also parallel to the axis of the parabola. K is the midpoint of FA and $HK = KC$. Take HC as a lever, or balance bar, with fulcrum at K. Place OP with its center at H, and leave OM where it is.

(a) Using the geometrical fact that $OM/OP = AC/AO$, show, by Archimedes' method of equilibrium, that the area of the parabolic segment is one-third the area of triangle AFC.

(b) Let U be the midpoint of AC and let the parallel through U to the axis of the parabola cut the parabola in V and FC in W. Using the geometrical fact that V lies on HC, prove, from part (a), that the area of the parabolic segment is four-thirds the area of triangle AVC.

The procedure of parts (a) and (b), Archimedes tells us in *Method*, is how he discovered the result in part (b). His mathematical conscience, however, would not permit him to accept the procedure as a proof. He accordingly supplied a rigorous demonstration, using the Eudoxian method of exhaustion, in his paper *Quadrature of the Parabola*.

9.7. Let us be given two curves m and n, and a point O. Suppose we permit ourselves to mark, on a given straightedge, a segment MN, and then to adjust the straightedge so that it passes through O and cuts the curves m and n with M on m and N on n. The line drawn along the straightedge is then said to have been drawn by "the insertion principle." Problems beyond the Euclidean tools can sometimes be solved with these tools if we also permit ourselves to use the insertion principle. Establish the correctness of the following

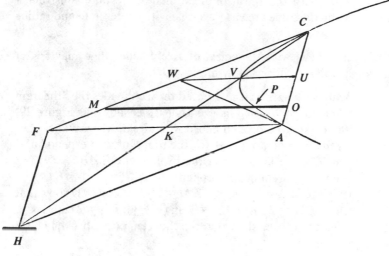

construction, using the insertion principle, for trisecting a general angle.

Let AOB be any central angle in a given circle. Through B draw a line BCD cutting the circle again in C, AO produced in D, and such that $CD = OA$, the radius of the circle. Then angle $ADB = \frac{1}{3}$ (angle AOB).

This solution of the trisection problem is implied by a theorem given by Archimedes.

9.8. A neat solution of the quadrature problem (that is, the problem of constructing a square equal in area to that of a given circle) can be achieved with the *spiral of Archimedes*, and we are told that Archimedes actually used his spiral for this purpose. We may define the spiral, in dynamical terms, as the locus of a point P moving uniformly along a ray which, in turn, is uniformly rotating in a plane about its origin. If we take for a polar frame of reference the position OA of the rotating ray when P coincides with the origin O of the ray, we have that OP is proportional to angle AOP, and the polar equation of the spiral is $r = a\theta$, a being the constant of proportionality.

Show how, with the spiral of Archimedes, we may construct a square having the same area as the circle with center at O and radius a.

9.9. Show how, with the spiral of Archimedes, one may trisect (more generally, multisect) any angle AOB.

9.10. Arabic scholars have attributed to Archimedes the "theorem of the broken chord," which asserts that if, as shown as Figure 29, AB and BC make up a broken chord in a circle, where $BC > AB$, and if M is the midpoint of arc ABC, the foot F of the perpendicular from M on BC is the midpoint of the broken chord ABC.

(a) Prove the theorem of the broken chord.

(b) Setting arc $MC = 2x$ and arc $BM = 2y$, successively show that $MC = 2\sin x$, $BM = 2\sin y$, $AB = 2\sin(x - y)$, $FC = 2\sin x \cos y$, $FB = 2\sin y \cos x$. Now show that the theorem of the broken chord yields the identity

$$\sin(x - y) = \sin x \cos y - \sin y \cos x.$$

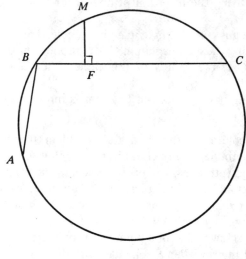

Fig. 29

(c) Using the theorem of the broken chord, obtain the identity

$$\sin(x + y) = \sin x \cos y + \sin y \cos x.$$

Further Reading

DIJKSTERHUIS, E. J., *Archimedes*. New York: Humanities Press, 1957.

HEATH, T. L., *The Works of Archimedes*. New York: Cambridge University Press, 1912. Reprinted by Dover, New York.

A BOOST FROM ASTRONOMY

The beginnings of trigonometry are obscure. So far as the pre-Hellenic period is concerned, there are some problems in the Rhind papyrus (ca. 1650 B.C.) that involve the cotangent of the dihedral angle at the base of a regular square pyramid, and there is the Babylonian cuneiform tablet known as Plimpton 322* (1900 to 1600 B.C.), which essentially contains a remarkable table of secants of fifteen angles ranging between 45° and 30°. It may well be that further studies into the mathematics of ancient Mesopotamia will disclose a substantial development of practical trigonometry. Babylonian astronomers had amassed a considerable collection of observational data, and we know that much of this information passed on to the Greeks. It was this early astronomy that gave birth to spherical trigonometry.

One of the earliest of the Greek astronomers was Aristarchus of Samos (ca. 310–230 B.C.), who is said to have applied mathematics to astronomy and to be the first to put forward the heliocentric theory of the solar system.† Nothing of his writings has come down to us, but it has been reported that, in his tract *On Sizes and Distances of the Sun and Moon,* he used the equivalent of the fact that

$$\frac{\sin a}{\sin b} < \frac{a}{b} < \frac{\tan a}{\tan b},$$

where $0 < b < a < \pi/2$.

---

*That is, the item with catalogue number 322 in the G. A. Plimpton archaeological collection at Columbia University.

†Aristarchus is sometimes referred to as the "Copernicus of antiquity."

The next eminent Greek mathematician-astronomer of antiquity known to us was Hipparchus, who was born in Nicaea of Asia Minor and who flourished around 140 B.C. Though Hipparchus reported an observation of the vernal equinox at Alexandria in 146 B.C., his most important astronomical observations were made at the famous observatory at Rhodes. Renowned as a careful and precise observer, he is credited with such accomplishments as the determination of the mean lunar month to within one second of the present accepted value, an accurate calculation of the inclination of the ecliptic, and the discovery and estimation of the annual precession of the equinoxes. He is said also to have computed the lunar parallax, to have determined the moon's perigee, and to have catalogued 850 fixed stars. He advocated the use of latitude and longitude to locate positions on the earth's surface, and he may have been the first to introduce into Greece the division of a circle into 360°. Though our knowledge of these achievements is only second-hand—inasmuch as almost nothing of Hipparchus' writings has come down to us—the implication is that Hipparchus was aware of the basic trigonometry of the celestial sphere.

A more direct, and very important, connection of Hipparchus with trigonometry is the crediting to Hipparchus, by the fourth-century commentator Theon of Alexandria, of a 12-book treatise dealing with the construction of a *table of chords*. This table is lost to us, but a subsequent table, given by Claudius Ptolemy (ca. 85–ca. 165)* and believed to have been adapted from Hipparchus' treatise, has survived. Ptolemy's table gives the lengths of the chords of all central angles of a given circle by half-degree intervals from ½° to 180°. The radius of the circle is divided into 60 equal parts and the chord lengths then expressed sexagesimally in terms of one of these parts as a unit. Thus, using the symbolism crd α to represent the length of the chord of a central angle α, one finds recordings like

$$\text{crd } 36° = 37^p \ 4' \ 55'',$$

meaning that the chord of a central angle of 36° is equal to 37/60 (or 37 small parts) of the radius, plus 4/60 of one of these small parts,

---

*Not to be confused with any of the erstwhile kings of Egypt bearing the name Ptolemy.

plus 55/3600 more of one of these small parts. It is evident from Figure 30 that a table of chords is equivalent to a table of trigonometric sines, for

$$\sin \alpha = \frac{AM}{OA} = \frac{AB}{\text{diameter of circle}} = \frac{\text{crd } 2\alpha}{120}.$$

Thus Ptolemy's table of chords gives, in reality, the sines of angles by quarter-degree intervals from 0° to 90°. There are reports that Hipparchus made systematic use of his table of chords and apparently was aware of the equivalent of several formulas now used in the solution of right spherical triangles.

Theon has also mentioned a six-book treatise on chords in a circle written by Menelaus of Alexandria (ca. 100), but this work also, along with a variety of others by Menelaus, is lost to us. There is, however, a three-book treatise by Menelaus, called *Sphaerica*, that has been preserved in the Arabic. This work throws considerable light on the Greek development of trigonometry.

The disappearance of so much of the early Greek work in astronomy is due to the fact that Ptolemy wrote a treatise that so

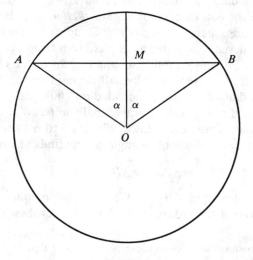

FIG. 30

eclipsed these earlier works that they were rendered superfluous. It was about A.D. 150 that Ptolemy wrote his great definitive Greek work on astronomy. This highly influential treatise, called the *Syntaxis mathematica,* or "Mathematical Collection," is remarkable for its completeness, compactness, and elegance. To distinguish it from other, lesser works on astronomy, later commentators assigned to it the superlative *magiste,* or "greatest." Still later, the Arabian translators prefixed the Arabian article *al,* and the work has ever since been known as the *Almagest.* The treatise is in thirteen books, and it is in Book I that we find, among some preliminary astronomical material, the table of chords referred to above, along with a succinct explanation of its derivation from a fertile geometrical proposition now known as *Ptolemy's theorem: In a cyclic quadrilateral the product of the diagonals is equal to the sum of the products of the two pairs of opposite sides.*

Certainly practical trigonometry cannot progress very far without the use of so-called trigonometric tables. The earliest systematic construction of an applicable trigonometric table therefore marks a GREAT MOMENT IN MATHEMATICS. It is the purpose of the rest of this lecture to give the gist of Ptolemy's method of constructing his highly useful table of chords, or table of sines.* For convenience of presentation and ease of understanding, we shall use modern algebraic notation, employ modern decimal fractions in place of the ancient sexagesimal fractions, and carry out the details of the development in small steps.

1. We start by establishing Ptolemy's theorem quoted above. To this end let $ABCD$ (see Figure 31) be any simple quadrilateral inscribed in a circle and let E be the point on the diagonal AC such that $\angle ABE = \angle DBC$. From the similar triangles ABE and DBC we have $AB/AE = DB/DC$, whence $(AB)(DC) = (DB)(AE)$. Again, from the similar triangles ABD and EBC we have $AD/DB = EC/CB$, whence $(AD)(CB) = (DB)(EC)$. It follows that

---

*It is likely that the method had been earlier employed by Hipparchus. The much briefer secant table found in Plimpton 322 appears to have been cleverly constructed from a collection of primitive Pythagorean triangles and is not nearly as applicable as Ptolemy's table.

$$(AB)(DC) + (AD)(CB) = DB(AE + EC) = (DB)(AC),$$

and the theorem is established.

We now establish three corollaries to Ptolemy's theorem.

2. COROLLARY 1. *If a and b are the chords of two arcs of a circle of unit radius, then*

$$s = (a/2)(4 - b^2)^{1/2} + (b/2)(4 - a^2)^{1/2}$$

is the chord of the sum of the two arcs.

Apply Ptolemy's theorem to the quadrilateral of Figure 32 where AC is a diameter, $BC = a$, and $CD = b$.

3. COROLLARY 2. *If a and b, a $\geq$ b, are the chords of two arcs of a circle of unit radius, then*

$$d = (a/2)(4 - b^2)^{1/2} - (b/2)(4 - a^2)^{1/2}$$

is the chord of the difference of the two arcs.

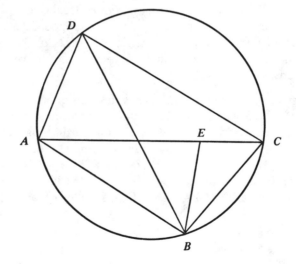

FIG. 31

Apply Ptolemy's theorem to the quadrilateral of Figure 33, where AB is a diameter, $BD = a$, and $BC = b$.

4. COROLLARY 3. *If t is the chord of a minor arc of a circle of unit radius, then*

$$h = [2 - (4 - t^2)^{1/2}]^{1/2}$$

is the chord of half the arc.

Apply Ptolemy's theorem to the quadrilateral of Figure 34 where AC is a diameter, $BD = t$ and BD is perpendicular to AC. We obtain

$$2t = 2h(4 - h^2)^{1/2},$$

whence, by squaring and rearranging the terms, we find

$$h^4 - 4h^2 + t^2 = 0.$$

Solving this as a quadratic in h^2 we obtain

$$h^2 = 2 \pm (4 - t^2)^{1/2}.$$

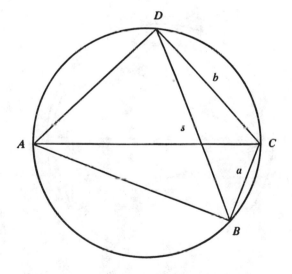

FIG. 32

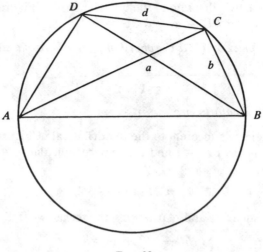

Fɪɢ. 33

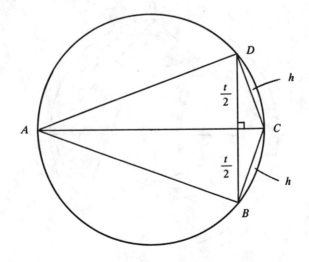

Fɪɢ. 34

Since *h* represents the chord of half the *minor* arc of the chord *BD*, we require the minus sign in the above result. Taking square roots we finally obtain

$$h = [2 - (4 - t^2)^{1/2}]^{1/2}.$$

5. Consider an isosceles triangle *AOB* (see Figure 35) with vertex angle *AOB* = 36°. Draw *AC* bisecting ∡*BAO*. Then from similar triangles *AOB* and *BAC* we have *AB/CB* = *OB/AB*. Setting *AB* = *x* and taking *OB* = 1, we find

$$x/(1 - x) = 1/x \quad \text{or} \quad x^2 + x - 1 = 0,$$

whence (to four-decimal-place accuracy)

$$x = (\sqrt{5} - 1)/2 = 0.6180.$$

It follows that in a circle of unit radius, crd 36° = 0.6180.*

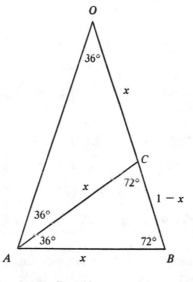

Fig. 35

---

*This is, of course, the *golden ratio* considered in LECTURE 5.

6. Since in a circle of unit radius, crd 60° = 1, we now find, by Corollary 2 above, that in the unit circle

$$\text{crd } 24° = \text{crd}(60° - 36°) = 0.4158.$$

7. By Corollary 3 we may now successively calculate the chords, in the unit circle, of 12°, 6°, 3°, 90′, and 45′, obtaining

$$\text{crd } 90′ = 0.0262 \quad \text{and} \quad \text{crd } 45′ = 0.0131.$$

8. By the relation

$$\frac{\sin a}{\sin b} < \frac{a}{b}, \qquad b < a < 90°,$$

the equivalent of which we pointed out early in our lecture was known to Aristarchus, we have

$$\text{crd } 60′/\text{crd } 45′ < 60/45 = 4/3,$$

or

$$\text{crd } 1° < (4/3)(0.0131) = 0.01747.$$

Also

$$\text{crd } 90′/\text{crd } 60′ < 90/60 = 3/2,$$

or

$$\text{crd } 1° > (2/3)(0.0262) = 0.01747.$$

It follows that, to four decimal place accuracy, crd 1° = 0.0175.

9. By Corollary 3 we may now find crd ½°.

10. Now one can construct a table of chords in the unit circle for ½° intervals.

Much of the subsequent work in practical trigonometry was the construction of ever better trigonometric tables. Thus the tenth-century Moslem mathematician Abû'l-Wefâ (940–998) computed a table of sines and tangents for 15′ intervals. Later, a table of sines was computed by the Viennese mathematician Georg von Peurbach (1423–1461) and a table of tangents by the German mathematician

Johann Müller* (1436–1476). George Joachim Rhaeticus (1514–1576), the leading Teutonic mathematical astronomer of the sixteenth century, spent 12 years with hired computers forming two remarkable and still useful trigonometric tables. One was a 10-place table of all six of the trigonometric functions, for every 10″ of arc; the other was a 15-place table for sines of every 10″ of arc, along with the first, second, and third differences. It is interesting that the Mathematics Advisory Board of the well-known *CRC Handbook of Tables for Mathematics* almost voted to delete the trigonometric tables from the *Handbook's* fifth edition; the wide proliferation of pocket calculators has rendered these tables rather superfluous.

We conclude with a brief word about the meanings of the present names of the trigonometric functions. With the exception of *sine*, the meanings are all clear from the geometrical interpretations of the functions when the angle is taken as a central angle of a circle of unit radius. Thus, in Figure 36, if the radius of the circle is one unit, the measures of tan α and sec α are given by the lengths of the tangent segment *CD* and the secant segment *OD*. And, of course, *cotangent* merely means "complement's tangent," and so on. The tangent, cotangent, secant, and cosecant functions have been known by various other names, these present ones not appearing until the end of the sixteenth century.

The origin of our word *sine* is deeper. The Hindu mathematician, Āryabhata the Elder (ca. 475–ca. 550), called it *ardhâ-jyā* ("half-chord") and also *jyā-ardhâ* ("chord-half"), and then abbreviated the term by simply using *jyā* ("chord"). From *jyā* the Arabs phonetically derived *jîba*, which, following the Arabian practice of omitting vowels, was written as *jb*. Now *jîba*, aside from its technical significance, is a meaningless word in Arabic. Later writers, coming across *jb*, as an abbreviation for the meaningless *jîba*, and knowing Arabic but no Sanskirt, substituted *jaib* instead, which contains the same letters and is a good Arabic word meaning "cove" or "bay." Still later, Gherardo of Cremona (ca. 1150), when he made his translations from the Arabic into Latin, replaced the Arabic *jaib* by its Latin equivalent *sinus*, whence came our present word *sine*.

---

*Better known as Regiomontanus, a Latinized form of his birthplace of Königsberg ("king's mountain").

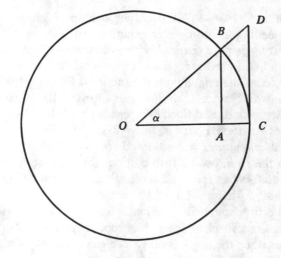

Fig. 36

Exercises

10.1. From a knowledge of the graphs of the functions sin x and tan x show that $(\sin x)/x$ decreases and $(\tan x)/x$ increases as x increases from 0 to $\pi/2$, and thus establish the inequalities

$$\frac{\sin a}{\sin b} < \frac{a}{b} < \frac{\tan a}{\tan b},$$

where $0 < b < a < \pi/2$.

10.2. Show that Corollaries 1, 2, and 3 of the lecture text are equivalent to the trigonometric identities

$$\sin(\alpha + \beta) = \sin \alpha \cos \beta + \cos \alpha \sin \beta,$$
$$\sin(\alpha - \beta) = \sin \alpha \cos \beta - \cos \alpha \sin \beta,$$
$$\sin(\theta/2) = [(1 - \cos \theta)/2]^{1/2},$$

where $0 < \alpha, \beta, \theta/2 < \pi/2$.

10.3. Establish the following consequences of Ptolemy's theorem: If P lies on the arc AB of the circumcircle of

(a) an equilateral triangle ABC, then $PC = PA + PB$.

(b) a square $ABCD$, then $(PA + PC)PC = (PB + PD)PD$.

(c) a regular pentagon $ABCDE$, then $PC + PE = PA + PB + PD$.

(d) a regular hexagon $ABCDEF$, then $PD + PE = PA + PB + PC + PF$.

10.4. A point lying on a side line of a triangle, but not coinciding with a vertex of the triangle, is called a *menelaus point* of the triangle for this side. Prove the following chain of theorems, wherein all segments and angles are directed (or sensed) segments and angles:

(a) *Menelaus' theorem.* A necessary and sufficient condition for three menelaus points D, E, F for the sides BC, CA, AB of a triangle ABC to be collinear is that

$$\left(\frac{BD}{DC}\right)\left(\frac{CE}{EA}\right)\left(\frac{AF}{FB}\right) = -1.$$

(b) If vertex O of a triangle BOC is joined to a point D (other than B or C) on line BC, then

$$\frac{BD}{DC} = \frac{OB \sin BOD}{OC \sin DOC}.$$

(c) Let D, E, F be menelaus points on the sides BC, CA, AB of a triangle ABC, and let O be a point in space not in the plane of triangle ABC. Then the points D, E, F are collinear if and only if

$$\left(\frac{\sin BOD}{\sin DOC}\right)\left(\frac{\sin COE}{\sin EOA}\right)\left(\frac{\sin AOF}{\sin FOB}\right) = -1.$$

(d) Let D', E', F' be three menelaus points on the sides $B'C'$, $C'A'$, $A'B'$ of a spherical triangle $A'B'C'$. Then D', E', F', lie on a great circle of the sphere if and only if

$$\left(\frac{\sin \widehat{B'D'}}{\sin \widehat{D'C'}}\right)\left(\frac{\sin \widehat{C'E'}}{\sin \widehat{E'A'}}\right)\left(\frac{\sin \widehat{A'F'}}{\sin \widehat{F'B'}}\right) = -1.$$

(This is the spherical case of the Menelaus theorem used by Menelaus in his *Sphaerica.*)

10.5. Establish the following chain of theorems:

(a) The product of two sides of a triangle is equal to the product of the altitude on the third side and the diameter of the circumscribed circle.

(b) Let *ABCD* be a cyclic quadrilateral of diameter t. Denote the lengths of sides *AB, BC, CD, DA* by a, b, c, d, the diagonals *BD* and *AC* by m and n, and the angle between either diagonal and the perpendicular upon the other by θ. Then

$$mt \cos \theta = ab + cd, \qquad nt \cos \theta = ad + bc.$$

(c) In the above quadrilateral

$$m^2 = \frac{(ac + bd)(ab + cd)}{ad + bc},$$

$$n^2 = \frac{(ac + bd)(ad + bc)}{ab + cd}.$$

(d) If, in the above quadrilateral, the diagonals are perpendicular to each other, then

$$t^2 = \frac{(ad + bc)(ab + cd)}{ac + bd}.$$

(e) *Ptolemy's second theorem.* In the above quadrilateral

$$\frac{n}{m} = \frac{ad + bc}{ab + cd}.$$

For those interested we here state an extension of Ptolemy's theorem and a singularly beautiful generalization of the theorem.

Extension of Ptolemy's theorem. In a convex quadrilateral *ABCD*,

$$(BC)(AD) + (CD)(AB) \geq (BD)(AC),$$

with equality if and only if the quadrilateral is cyclic.

Generalization of Ptolemy's theorem. Let $T_1 T_2 T_3 T_4$ be a convex quadrilateral inscribed in a circle C. Let C_1, C_2, C_3, C_4 be four circles touching circle C externally at T_1, T_2, T_3, T_4, respectively. Then

$$t_{12} t_{34} + t_{23} t_{41} = t_{13} t_{24},$$

where t_{ij} is the length of a common external tangent to circles C_i and C_j. [This is a special case of a more general theorem due to John Casey (1820–1891).]

Further Reading

AABOE, ASGER, *Episodes from the Early History of Mathematics* (New Mathematical Library, No. 13). New York: Random House and L. W. Singer, 1964. (The New Mathematical Library became a publication series of the Mathematical Association of America, Washington, D.C., in 1975.)

HEATH, T. L., *History of Greek Mathematics,* 2 vols. New York: Oxford University Press, 1931.

THE FIRST GREAT NUMBER THEORIST

There are two aspects to number study—the search for relationships among numbers and the development of the art of computing with numbers. To the ancient Greeks the former was known as *arithmetic* and the latter as *logistic*. This classification persisted through the Middle Ages until around the close of the fifteenth century, when texts appeared treating both aspects of number work under the single name *arithmetic*. It is interesting that today *arithmetic* has its original meaning in continental Europe, while in England and the United States the popular meaning of *arithmetic* is synonymous with that of ancient *logistic*, and in these two countries the theoretical side of number study is designated by the descriptive term *number theory*.

It is generally agreed that the first stimulating steps in the development of number theory were taken by Pythagoras and his followers, in conjunction with the philosophy of the Pythagorean Brotherhood that the whole numbers control the universe. Much of this work became the basis for future number mysticism. Thus Iamblichus, an influential Neoplatonic philosopher of about A.D. 320, has ascribed to Pythagoras the discovery of *amicable*, or *friendly, numbers*. Two positive integers are *amicable* if each is the sum of the proper divisors* of the other. Thus 284 and 220, the pair ascribed to Pythagoras, are amicable since the proper divisors of 220, namely, 1, 2, 4, 5, 10, 11, 20, 22, 44, 55, and 110, sum to 284, and the proper divisors of 284, namely, 1, 2, 4, 71,

---

*The *proper divisors* of a positive integer n are all the positive integral divisors of n except n itself. Note that 1 is a proper divisor of n. A somewhat antiquated synonym for proper divisor is *aliquot part*.

and 142, sum to 220. This pair of numbers attained a mystical aura, and superstition later maintained that two talismans bearing these numbers would seal perfect friendship between the wearers. The numbers came to play an important role in magic, sorcery, astrology, and the casting of horoscopes.

After the original pair, 284 and 220, of amicable numbers, no new ones were found until the great French number theorist Pierre de Fermat in 1636 announced 17,296 and 18,416 as another pair. Two years later the French mathematician and philosopher René Descartes gave a third pair. The Swiss mathematician Leonard Euler undertook a systematic search for amicable numbers and, in 1747, gave a list of 30 pairs, which he later extended to more than 60. A curiosity in the history of these numbers was the late discovery, by the sixteen-year-old Italian boy Nicolo Paganini in 1866, of the overlooked and relatively small pair 1184 and 1210. Today more than 1,000 pairs of amicable numbers are known.

The concept of amicable numbers has led to some generalizations in modern times. For example, a cyclic sequence of three or more numbers such that the sum of the proper divisors of each is equal to the next in the sequence is known as a *sociable chain* of numbers. Only two sociable chains involving numbers below 1,000,000 are known, one of five "links" (found by the Frenchman P. Poulet) starting with 12,496 and one of 28 links starting with 14,316.* A sociable chain of exactly three links is called a *crowd*; no crowds have yet been found.

Other numbers having properties mystically employed in numerological speculations, and generally ascribed to the Pythagoreans, are the *perfect, deficient,* and *abundant numbers.* Let N represent the sum of the proper divisors of a positive integer n. Then n is said to be *perfect, deficient,* or *abundant* according as $N = n$, $N < n$, or $N > n$. Thus 6 (with proper divisors 1, 2, 3) is perfect, 8 (with proper divisors 1, 2, 4) is deficient, and 12 (with proper divisors 1, 2, 3, 4, 6) is abundant.

Until 1952, there were only 12 known perfect numbers, all of them even numbers, of which the first three are 6, 28, and 496. The last proposition of the ninth book of Euclid's *Elements* proves that *if*

---

*There are some four-link chains known involving numbers above 1,000,000.

$2^n - 1$ *is a prime number,* * *then* $2^{n-1}(2^n - 1)$ *is a perfect number.* The perfect numbers given by Euclid's formula are even numbers, and Euler has shown that every even perfect number must be of this form. The existence or nonexistence of odd perfect numbers is one of the celebrated unsolved problems in number theory; it is known that there is no such number less than 10^{100}.

In 1952, with the aid of the SWAC digital computer, five more perfect numbers were discovered, corresponding to $n = 521$, 607, 1279, 2203, and 2281 in Euclid's formula. In 1957, using the Swedish machine BESK, another was found, corresponding to $n = 3217$, and in 1961, with an IBM 7090, two more were found, for $n = 4253$ and 4423. There are no other even perfect numbers for $n < 5000$. The values $n = 9,689, 9,941, 11,213, 19,937, 21,701, 23,209$, and 44,497 have been found to yield further perfect numbers, bringing the list of known perfect numbers to 27.

The concept of perfect numbers has inspired certain generalizations by modern mathematicians. If we let $\sigma(n)$ represent the sum of *all* the divisors of n (including n itself), then n is perfect if and only if $\sigma(n) = 2n$. In general, if we should have $\sigma(n) = kn$, where k is a natural number, then n is said to be *k-tuply perfect.* One can show, for example, that 120 and 672 are triply perfect. It is not known if there exist infinitely many multiply perfect numbers, let alone just perfect ones. Nor is it known if there exists any odd multiply perfect number. In 1944 the concept of *superabundant numbers* was created. A natural number n is *superabundant* if and only if $\sigma(n)/n > \sigma(k)/k$ for all $k < n$. It is known that there are infinitely many superabundant numbers. Other numbers related to perfect, deficient, and abundant numbers that have been introduced in recent times are *practical numbers, quasiperfect numbers, semiperfect numbers,* and *weird numbers.* We merely mention these concepts to illustrate how ancient number work has inspired related modern investigations.

While we are not absolutely certain that amicable, perfect, de-

---

*A *prime number* is a positive integer greater than 1 and having no positive integral divisors other than itself and unity. An integer greater than 1 that is not a prime number is called a *composite number.* Thus 7 is a prime number, whereas 12 is a composite number.

ficient, and abundant numbers originated with the Pythagoreans, there seems no doubt that the so-called *polygonal numbers* arose with the earliest members of the Pythagorean society. These numbers, considered as the number of dots in certain geometrical configurations, represent a link between geometry and arithmetic. Figures 37, 38, and 39 account for the geometrical nomenclature of *triangular numbers, square numbers, pentagonal numbers,* and so on. There are many pretty theorems about polygonal numbers which can be established very simply from the pictorial representations of these numbers. Consider, for example:

1. *Any square number is the sum of two successive triangular numbers.* (See Figure 40.)

2. *The nth pentagonal number is equal to n plus three times the $(n-1)th$ triangular number.* (See Figure 41.)

3. *The sum of any number of consecutive odd integers, starting with 1, is a square number.* (See Figure 42.)

Of course these theorems can also be established in purely algebraic fashion, and there are deeper properties of the polygonal numbers where an algebraic treatment cannot easily be circumvented.

The *prime numbers,* as the building bricks from which all other integers are multiplicatively made, have enjoyed a long history, running from the days of the ancient Greeks up into present times. Euclid, in Proposition 20 of Book IX of his *Elements*, proved that the set of prime numbers is infinite. A beautiful generalization of

Triangular numbers

 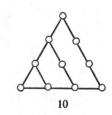

and so on

1   3   6   10

FIG. 37

Square numbers

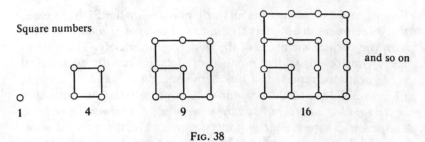

and so on

1   4   9   16

FIG. 38

Pentagonal numbers

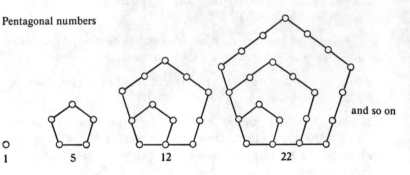

and so on

1   5   12   22

FIG. 39

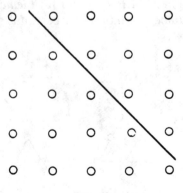

FIG. 40

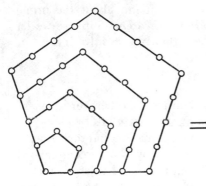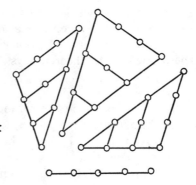

FIG. 41

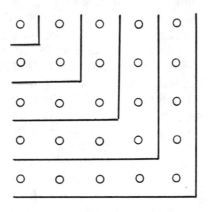

FIG. 42

this theorem was established by Peter Gustav Lejeune Dirichlet (1805–1859), who succeeded in showing that every arithmetic sequence

$$a, a + d, a + 2d, a + 3d, \ldots,$$

in which a and d are relatively prime, contains infinitely many primes. The proof of this result is far from simple.

Probably the most amazing result yet found about the prime numbers is the so-called *prime number theorem*. Suppose we let A_n denote the number of primes less than the positive integer n. The prime number theorem says that

$$(A_n \log_e n)/n$$

approaches 1 as n increases without bounds. In other words, A_n/n, called the *density* of the primes among the first n positive integers, is approximated by $1/\log_e n$, the approximation improving as n increases. This theorem was conjectured by the fifteen-year-old C. F. Gauss (1777–1855) from an examination of a large table of primes, and was independently established in 1896 by the French and Belgian mathematicians J. Hadamard and C. J. de la Vallée Poussin.

Extensive factor tables are invaluable for research on prime numbers. Such a table for all numbers up to 24,000 was published by J. H. Rahn in 1659, as an appendix to a book on algebra. In 1668, John Pell of England extended this table up to 100,000. As a result of appeals by the German mathematician J. H. Lambert, an extensive and ill-fated factor table was computed by a Viennese schoolmaster named Antonio Felkel. The first volume of Felkel's computations, giving factors of numbers up to 408,000, was published in 1776 at the expense of the Austrian imperial treasury. But, as there were very few subscribers to the volume, the treasury recalled almost the entire edition and converted the paper into cartridges to be used in a war for killing Turks. In the nineteenth century, the combined efforts of Chernac, Burckhardt, Crelle, Glaisher, and the lightning mental calculator Dase, led to a factor table covering numbers up to 10,000,000 and published in ten volumes. The greatest achievement of this sort, however, is the table calculated by J. P. Kulik (1773–1863), of the University of Prague. His as yet unpublished manuscript is the result of a 20-year hobby, and covers all numbers up to 100,000,000. The best available factor table is that of the American mathematician D. N. Lehmer (1867–1938); it is a cleverly assembled one-volume table covering numbers up to 10,000,000. Lehmer has pointed out that Kulik's table contains errors.

There are enough unsettled questions about prime numbers to

fill a booklet of their own. For example: Are there infinitely many primes of the form $n^2 + 1$? Is there always a prime between n^2 and $(n + 1)^2$? Is any integer n from some point onward either a square or the sum of a prime and a square? Are there infinitely many *Fermat primes*—that is, primes of the form $2^{2^n} + 1$?

As one further set of numbers that received attention by the early Greeks, we might mention *Pythagorean triples*. A *Pythagorean triple* is a trio (a, b, c) of positive integers such that $a^2 + b^2 = c^2$. It follows that a, b, and c can serve as lengths of the legs and the hypotenuse of a right triangle. Such a right triangle is called a *Pythagorean triangle*. Thus $(3, 4, 5)$ and $(5, 12, 13)$ are Pythagorean triples, yielding the 3-4-5 and the 5-12-13 Pythagorean triangles. It has been found that the ancient Babylonians prior to 1600 B.C. were acquainted with a method of finding Pythagorean triples. A vast research literature has been developed concerning these number triples.

Now in the history of mathematics there is one man who stands out as probably the first true genius in the field of number theory, and one of whose works so profoundly influenced later European number theorists that the production of this work can well be labeled a GREAT MOMENT IN MATHEMATICS. The man is Diophantus of Alexandria, and the work alluded to is his famous *Arithmetica*. Though there is some tenuous evidence placing Diophantus in the first century of our era, most historians tend to put him in the third century. Beyond the fact that he flourished at Alexandria, nothing certain is known about his personal life.

Diophantus wrote three mathematical works: *Arithmetica*, of which six of the original thirteen books are extant, *On Polygonal Numbers*, of which only a fragment exists, and *Porisms*, which is lost.

The *Arithmetica* is a great and highly original work. It is an analytical treatment of algebraic number theory that marks the author as a cunning virtuoso in this field. The work has had many commentators, but it was Regiomontanus who, in 1463, called for a Latin translation of the extant Greek text. The challenge was taken up in 1575 by Xylander (the Greek name assumed by Wilhelm Holzmann, a professor at the University of Heidelberg), who made a very meritorious translation accompanied by valuable commentary. The Xylander translation was used, in turn, by the Frenchman Bachet de Méziriac, who in 1621 published the first edition of the Greek text along with a Latin translation and notes. A second edition, unfortunately carelessly printed, of the de Méziriac translation was brought out in 1670. This second edition is especially important historically because it contained, incorporated into the text, Pierre de Fermat's famous marginal notes that stimulated extensive research in number theory. French, German, and English translations of the *Arithmetica* appeared later.

The extant portion of the *Arithmetica* is devoted to the solution of about 130 problems, of considerable variety, leading to equations of the first and second degree; one very special cubic equation is solved. The first book concerns itself with determinate equations in one unknown, and the other books with indeterminate equations of the second degree in two and three unknowns. Striking is the lack of general methods, but rather the invention of clever mathematical devices designed for the needs of each individual problem. Diophantus recognized only positive rational answers and was, in most cases, satisfied when he found one answer to a problem, though many different answers might exist.

There are some deep number theorems stated in the *Arithmetica*. For example, we find, without a proof but with an allusion to *Porisms*, that the difference of two rational cubes is also the sum of two rational cubes—a matter that was later investigated by François Viète, Bachet de Méziriac, and Pierre de Fermat. There are many propositions concerning the representation of numbers as the sum of two, three, or four squares, a field of investigation later completed by Pierre de Fermat, Leonard Euler, and Joseph Louis Lagrange. It might be interesting to list a few of the problems found in the *Arithmetica*; they are all beguiling and many of them are challenging. It must be borne in mind that by "number" is meant "positive rational number."

Problem 17,* Book I: Find four numbers, the sum of every arrangement three at a time being given; say, 22, 24, 27, and 20.

---

*The numbering of the problems is that assigned to them in T. L. Heath's *Diophantus of Alexandria*, 2nd ed.

Problem 28, Book II: Find two square numbers such that their product added to either gives a square number. (Diophantus' answer: $(3/4)^2$, $(7/24)^2$.)

Problem 6, Book III: Find three numbers such that their sum is a square and the sum of any pair is a square. (Diophantus' answer: 80, 320, 41.)

Problem 7, Book III: Find three numbers in arithmetic progression such that the sum of any pair is a square. (Diophantus' answer: 120½, 840½, 1560½.)

Problem 13, Book III: Find three numbers such that the product of any two added to the third is a square.

Problem 15, Book III: Find three numbers such that the product of any two added to the sum of these two is a square.

Problem 10, Book IV: Find two numbers such that their sum is equal to the sum of their cubes. (Diophantus' answer: 5/7, 8/7.)

Problem 21, Book IV: Find three numbers in geometric progression such that the difference of any two is a square number. (Diophantus' answer: 81/7, 144/7, 256/7.)

Problem 1, Book VI: Find a Pythagorean triangle in which the hypotenuse minus each of the legs is a cube. (Diophantus' answer: 40-96-104.)

Problem 16, Book VI: Find a Pythagorean triangle in which the length of the bisector of one of the acute angles is rational.

Indeterminate algebraic problems in which one is to find only the rational solutions have become known as *Diophantine problems*. In fact, modern usage of the terminology usually implies the restriction of the solutions to integers. It is only fair to point out that Diophantus did not originate problems of this sort, but he did possess uncommon talent in dealing with such problems.

We close our lecture by remarking upon what has become perhaps the most famous of all Diophantine problems. Problem 8 of Book II of the *Arithmetica* reads: "To divide a given square number into two squares." In his copy of Bachet's translation of the *Arithmetica*, Fermat penned the following tantalizing marginal note: "To divide a cube into two cubes, a fourth, or in general any power whatever into two powers of the same denomination above the second is impossible, and I have assuredly found an admirable proof of this, but the margin is too narrow to contain it." In other words, Fermat claimed to have had a proof that *there do not exist positive integers x, y, z, n such that $x^n + y^n = z^n$, where n > 2*. This italicized statement has become known as *Fermat's last "theorem,"* and whether Fermat really possessed a sound demonstration of it will probably forever remain an enigma. Many of the most prominent mathematicians since Fermat's time have tried their skill on the problem, but the general statement still remains intractable. There is a proof given elsewhere by Fermat for the case $n = 4$, and Euler supplied a proof (later perfected by others) for $n = 3$. About 1825, independent proofs for the case $n = 5$ were given by Legendre and Dirichlet, and in 1839, Lamé proved the theorem for $n = 7$. Very significant advances in the study of the problem were made by the German mathematician E. Kummer (1810–1893). In 1843, Kummer submitted a purported proof to Dirichlet, who pointed out an error in the reasoning. Kummer then returned to the problem with renewed vigor, and a few years later, after developing an important allied subject in higher algebra called the *theory of ideals,* derived very general conditions for the insolubility of the Fermat relation. Almost all important subsequent progress on the problem has been based on Kummer's investigations. It is now known that Fermat's last "theorem" is certainly true for all $n < 100,000$, and for many other special values of n. In 1908, the German mathematician Paul Wolfskehl bequeated 100,000 marks to the Academy of Science at Göttingen as a prize for the first complete proof of the "theorem." The result was a deluge of alleged proofs by glory-seeking and money-seeking laymen, and ever since then the problem has haunted amateurs somewhat as does the trisection of an arbitrary angle and the squaring of the circle. Fermat's last "theorem" has the peculiar distinction of being the mathematical problem for which the greatest number of incorrect proofs have been published.

Exercises

11.1. (a) Show that Nicolo Paganini's numbers, 1184 and 1210, are amicable.

(b) T̲âbit ibn Qorra (826–901) invented the following rule for finding amicable numbers: *If*

$$p = (3)(2^n) - 1, \qquad q = (3)(2^{n-1}) - 1, \qquad r = (9)(2^{2n-1}) - 1$$

are three odd primes, then $2^n pq$ and $2^n r$ are a pair of amicable numbers. Verify this for $n = 2$ and $n = 4$.

11.2. (a) Show that in Euclid's formula for perfect numbers, n must be prime.

(b) What is the fourth perfect number furnished by Euclid's formula?

(c) Prove that the sum of the reciprocals of *all* the divisors of a perfect number is equal to 2.

11.3. (a) Show that if p is a prime, then p^n is deficient.

(b) Find the 21 abundant numbers less than 100. It will be noticed that they are all even numbers. To show that not all abundant numbers are even, show that $945 = 3^3 \cdot 5 \cdot 7$ is abundant. This is the first odd abundant number.

(c) Show that any multiple of an abundant or perfect number is abundant.

11.4. (a) List the first four hexagonal numbers.

(b) Show that the nth triangular number and the nth pentagonal number are given by

$$T_n = n(n + 1)/2, \quad P_n = n(3n - 1)/2.$$

(c) Supply algebraic demonstrations of the three theorems on polygonal numbers given in the lecture text.

(d) An *oblong number* is the number of dots in a rectangular array having one more column than rows. Show, geometrically and algebraically, that the sum of the first n positive even integers is an oblong number.

(e) Show, both geometrically and algebraically, that any oblong number is twice a triangular number.

(f) Show, geometrically and algebraically, that 8 times any triangular number, plus 1, is a square number.

(g) Prove that every even perfect number is also a triangular number.

(h) Prove that the sequence of m-gonal numbers is given by

$$an^2 + bn, \qquad n = 1, 2, \ldots,$$

for a certain fixed pair of rational numbers a and b.

(i) Find a and b of part (h) when $m = 7$.

11.5. (a) Eratosthenes (ca. 230 B.C.) is noted in arithmetic for the following device, known as the *sieve*, for finding all the prime numbers less than a given number n. One writes down, in order and starting with 3, all the odd numbers less than n. The composite numbers in the sequence are then sifted out by crossing off, from 3, every third number, then from the next remaining number, 5, every fifth number, then from the next remaining number, 7, every seventh number, from the next remaining number, 11, every eleventh number, and so on. In the process some numbers will be crossed off more than once. All the remaining numbers, along with the number 2, constitute the list of primes less than n.

Find, by the sieve of Eratosthenes, all the primes below 500.

(b) Prove that a positive integer p is prime if it has no prime factor not exceeding the greatest integer whose square does not exceed p. This theorem says that, in the elimination process of the sieve of Eratosthenes, we may stop as soon as we reach a prime $p > \sqrt{n}$, for the cancellation of every pth number from p will merely be a repetition of cancellations already effected. Thus, in finding the prime less than 500, we may stop after crossing off every nineteenth number from 19, since the next prime, 23, is greater than $\sqrt{500}$.

(c) Compute $(A_n \log_e n)/n$ for $n = 500$, 10^8, and 10^9.

(d) Prove that there can always be found n consecutive composite integers, however great n may be.

11.6. (a) Show that for any positive integer m, the three numbers

$$2m, \qquad m^2 - 1, \qquad m^2 + 1$$

constitute a Pythagorean triple. The Pythagoreans have been credited with knowledge of this.

(b) Prove that there exists no isosceles right triangle whose sides are integers.

(c) Prove that no Pythagorean triple exists in which one integer is a mean proportional between the other two.

If the three numbers of a Pythagorean triple contain no common positive integral factor other than unity, the triple is called a *primitive Pythagorean triple*. Thus (3,4,5) is a primitive triple, whereas (6,8,10) is not. It can be shown that all primitive Pythagorean triples (a,b,c) are given parametrically by

$$a = 2uv, \qquad b = u^2 - v^2, \qquad c = u^2 + v^2,$$

where u and v are relatively prime, one even and one odd, and $u > v$. Thus, if $u = 2$ and $v = 1$, we obtain the primitive triple $a = 4$, $b = 3$, $c = 5$. Find the 16 primitive Pythagorean triples (a,b,c) for which b is even and $c < 100$. Now show that there are exactly 100 distinct Pythagorean triples (a,b,c) with $c < 100$.

   (e) Show that if $(a,a + 1,c)$ is a Pythagorean triple, so also is

$$(3a + 2c + 1, 3a + 2c + 2, 4a \mid 3c \mid 2).$$

It follows that from a given Pythagorean triple whose legs are successive natural numbers, we can obtain another such Pythagorean triple with bigger sides.

   (f) Starting with the Pythagorean triple (3,4,5), find five more Pythagorean triples whose legs are successive natural numbers and whose sides are progressively bigger.

   (g) Prove that for any natural number $n > 2$ there exists a Pythagorean triple with a leg equal to n.

   (h) Prove that there are only a finite number of Pythagorean triples having a given leg a.

11.7. (a) Solve Problem 17 of Book I of the *Arithmetica*.

   (b) In the right triangle ABC, right-angled at C, AD bisects angle A. Solve Problem 16 of Book VI of the *Arithmetica* by finding the set of smallest integers for AB, AD, AC, BD, DC such that $DC : CA : AD = 3 : 4 : 5$.

   (c) If m is any positive integer and

$$x = m^2, \qquad y = (m + 1)^2, \qquad z = 2(x + y + 1),$$

show that the six numbers

$$xy + x + y, \qquad yz + y + z, \qquad zx + z + x,$$

$$xy + z, \qquad yz + x, \qquad zx + y$$

are all square numbers. Show that this solves Problems 13 and 15 of Book III of the *Arithmetica*.

(d) Augustus De Morgan, who lived in the nineteenth century, was fond of proposing the conundrum: "I was x years old in the year x^2." When was De Morgan born?

11.8. The early Hindus solved the problem of finding all integral solutions of the linear indeterminate equation $ax + by = c$, where a, b, c are integers.

(a) If $ax + by = c$ has an integral solution, show that the greatest common divisor of a and b is a divisor of c. (This theorem says that there is no loss in generality if we assume a and b to be relatively prime.)

(b) If x_1 and y_1 constitute an integral solution of $ax + by = c$, where a and b are relatively prime, show that all integral solutions are given by $x = x_1 + mb$, $y = y_1 - ma$, where m is an arbitrary integer. (This theorem says that all integral solutions are known if just one integral solution can be found.)

(c) Solve $7x + 16y = 209$ for positive integral solutions.

(d) Solve $23x + 37y = 3000$ for positive integral solutions.

(e) In how many ways can the sum of five dollars be paid in dimes and quarters?

11.9. Find the smallest permissible answer to the following indeterminate problem of Mahāvira (ca. 850): "Into the bright and refreshing outskirts of a forest, which was full of numerous trees with their branches bent down with the weight of flowers and fruits, trees such as jambu trees, lime trees, plantains, areco palms, jack trees, date palms, hintala trees, palmyras, punnāgo trees, and mango trees—outskirts, the various quarters whereof were filled with many sounds of crowds of parrots and cuckoos found near springs containing lotuses with bees roaming about them— into such forest outskirts a number of weary travelers entered with joy. There were 63 numerically equal heaps of plantain fruits put together and combined with 7 more of those same fruits, and these were equally distributed among 23 travelers so as to have no remainder. You tell me now the numerical measure of a heap of plantains."

11.10. (a) Show that to establish Fermat's last "theorem" it is sufficient to consider only prime exponents $p > 2$.

(b) Assuming Fermat's last "theorem," show that the curve $x^n + y^n = 1$, where n is a positive integer greater than 2, contains no points with rational coordinates except those points where the curve crosses a coordinate axis.

(c) Fermat proved that the area of an integral-sided right triangle cannot be a square number. Assuming this, show that the equation $x^4 - y^4 = z^2$ has no solution in positive integers x, y, z, and then prove Fermat's last "theorem" for the case $n = 4$.

Further Reading

HEATH, T. L., *Diophantus of Alexandria*, rev. ed. New York: Cambridge University Press, 1910.

SIERPIŃSKI, WACLAW, *A Selection of Problems In the Theory of Numbers*, tr. by A. Sharma. New York: Pergamon Press, 1964.

SIERPIŃSKI, WACLAW, *Pythagorean Triangles*, tr. by A. Sharma. New York: Yeshiva University, 1962.

THE SYNCOPATION OF ALGEBRA

In contrast to his geometry course, a high school student finds algebra a highly symbolized field of mathematical study. The work bristles with plus signs, minus signs, division symbolism, later letters of the alphabet for unknown quantities and early letters for fixed quantities, various symbols (parentheses, brackets, braces) for signs of aggregation, exponents, subscripts, radicals, equality signs, factorial symbols, combination and permutation symbolism, logarithmic notation, and so on. It is seldom realized by the student that all this symbolism is little more than four hundred years old— indeed, much of it is considerably less than four hundred years old.

It was G. H. F. Nesselmann who, in 1842, first characterized three stages in the historical evolution of algebraic symbolism. At the start there is *rhetorical algebra*, in which the solution of a problem is written, without any abbreviations or symbolism, as a pure prose argument. Then comes *syncopated algebra*, in which stenographic abbreviations are adopted for some of the more frequently recurring quantities, relations, and operations. Finally, as the last stage, we have *symbolic algebra*, in which solutions to problems appear largely in a mathematical shorthand made up of symbols having little apparent connections with the entities and ideas they represent.

It is probably quite correct to say that all algebra prior to the time of Diophantus of Alexandria (A.D. 250?) was rhetorical. One of Diophantus' significant contributions to algebraic development was his syncopation of Greek algebra. It must be confessed, however, that rhetorical algebra persisted quite generally in the rest of the world, with the exception, as we shall see, of India, for many hundreds of years. Specifically, in Western Europe algebra remained essentially rhetorical until the fifteenth century, when spotty instances of syncopation began to appear. Symbolic algebra made its first appearance in Western Europe in the sixteenth century, but developed so slowly that it did not become widespread until about the middle of the seventeenth century.

Perhaps our best source of ancient Greek algebra problems is a collection known as the *Palatine*, or *Greek*, *Anthology*, a group of 46 number problems stated in epigrammatic form, assembled about A.D. 500 by the grammarian Metrodorus. It is possible that some of these problems may have originated with the author, but there is every reason to believe that most of them are of considerably more ancient origin. The problems, apparently intended for mental recreation, are of a type alluded to by Plato (ca. 400 B.C.), and closely resemble some of the problems found in the Rhind papyrus (ca. 1650 B.C.). Half of the problems lead to simple linear equations in one unknown, a dozen more to easy simultaneous linear equations in two unknowns, one to three linear equations in three unknowns, one to four linear equations in four unknowns, and there are two instances of indeterminate linear equations. Many of the problems are very much like standard types found in our present-day algebra textbooks under the descriptive titles of "distribution" problems, "work" problems, "cistern" problems, "mixture" problems, "age" problems, and so on. Although in general the problems present little difficulty when attacked with modern algebraic symbolism, it must be conceded that rhetorical solutions would demand close mental attention. Let any disbeliever try his skill at purely rhetorical solutions of the following examples chosen somewhat randomly from the *Palatine Anthology*:

1. [a "distribution" problem] How many apples are needed if four persons of six receive one-third, one-eighth, one-fourth, and one-fifth, respectively, of the total number, while the fifth receives ten apples, and one apple remains left for the sixth person?

2. [an "age" problem] Demochares has lived a fourth of his life as a boy, a fifth as a youth, a third as a man, and has spent 13 years in his dotage. How old is he?

3. [a "work" problem] Brickmaker, I am in a hurry to erect this house. Today is cloudless, and I do not require many more bricks,

for I have all I want but three hundred. Thou alone in one day couldst make as many, but thy son left off working when he had finished two hundred, and thy son-in-law when he had made two hundred and fifty. Working all together, in how many days can you make these?

4. [a "cistern" problem] I am a brazen lion; my spouts are my two eyes, my mouth, and the flat of my right foot. My right eye fills a jar in two days (1 day = 12 hours), my left eye in three, and my foot in four. My mouth is capable of filling it in six hours. Tell me how long all four together will take to fill it.

5. [a "mixture" problem] Make a crown of gold, copper, tin, and iron weighing 60 minae: gold and copper shall be two thirds of it; gold and tin three fourths of it; and gold and iron three fifths of it. Find the weights of gold, copper, tin, and iron required.

The production of Diophantus' *Arithmetica* merits classification as a GREAT MOMENT IN MATHEMATICS not only, as we have seen in the preceding lecture, because of its remarkable and influential arithmetic content, but also, as we will now show, because in this work we find the first steps taken toward an algebraic notation. These steps were in the nature of stenographic abbreviations.

In the *Arithmetica* we find abbreviations for the unknown, powers of the unknown up through the sixth, subtraction, equality, and reciprocals. Our word "arithmetic" comes from the Greek word *arithmetike*, a compound of the words *arithmos* for "number" and *techne* for "science." T. L. Heath has rather convincingly pointed out that Diophantus' symbol for the unknown probably originated by merging the first two Greek letters, α and ρ, of the word *arithmos*. In time, this came to look like the Greek final sigma ς. While this is conjecture, there is no doubt about the meaning of the notation for powers of the unknown. Thus "unknown squared" is denoted by Δ^T, the first two letters of the Greek word *dunamis* ($\Delta \Upsilon N A M I \Sigma$) for "power." Again, "unknown cubed" is denoted by K^T, the first two letters of the Greek word *kubos* ($K \Upsilon B O \Sigma$) for "cube." Explanations are easily supplied for the succeeding powers of the unknown, $\Delta^T\Delta$ (square-square), ΔK^T (square-cube), and $K^T K$ (cube-cube). Diophantus' symbol for "minus" looks like an inverted V with the angle bisector drawn in. This has been explained as a compound of Λ and I, letters in the Greek word *leipis* (ΛEIΨIΣ) for "lacking." Addition is indicated by juxtaposition, and all negative terms in an expression are gathered together and preceded by the minus symbol. A numerical coefficient of any power of the unknown is represented by the appropriate alphabetic Greek numeral following the power symbol. If there is a constant term then $\overset{o}{M}$, an abbreviation of the Greek word *monades* (MONAΔEΣ) for "units," is used, with the appropriate numerical coefficient. The alphabetic Greek numerals are as follows:

1	α	alpha	10	ι	iota		100	ρ	rho	
2	β	beta	20	κ	kappa		200	σ	sigma	
3	γ	gamma	30	λ	lambda		300	τ	tau	
4	δ	delta	40	μ	mu		400	υ	upsilon	
5	ϵ	epsilon	50	ν	nu		500	ϕ	phi	
6	obsolete di-		60	ξ	xi		600	χ	chi	
	gamma		70	o	omicron		700	ψ	psi	
7	ζ	zeta	80	π	pi		800	ω	omega	
8	η	eta	90	obsolete koppa			900	obsolete sampi		
9	θ	theta								

As examples of the use of these numeral symbols we have

$$13 = \iota\gamma, \qquad 31 = \lambda\alpha, \qquad 742 = \psi\mu\beta.$$

Accompanying bars or accents were used for larger numbers. The symbols for the obsolete digamma, koppa, and sampi, respectively, are shown in Figure 43.

With the numerals and stenographic abbreviations above,

$$x^3 + 13x^2 + 8x \quad \text{and} \quad x^3 - 8x^2 + 2x - 3$$

would appear as

$$K^T\alpha\Delta^T\iota\gamma\,\varsigma\,\eta \quad \text{and} \quad K^T\alpha\,\varsigma\,\beta\Lambda\Delta^T\eta\overset{o}{M}\gamma,$$

which can be read literally as

unknown cubed 1, unknown squared 13, unknown 8

and

(unknown cubed 1, unknown 2) minus (unknown squared 8, units 3).

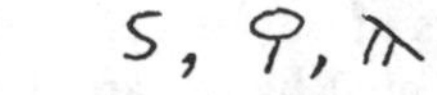

$$S, \; 9, \; \pi$$

Fig. 43

It was thus that rhetorical algebra first became syncopated algebra, and a GREAT MOMENT IN MATHEMATICS occurred.

As remarked earlier in this lecture, the Hindus also syncopated their algebra. As with Diophantus, addition was usually indicated by juxtaposition. Subtraction was indicated by placing a dot over the subtrahend, multiplication by writing *bha* (the first syllable of the word *bhavita*, "the product") after the factors, division by writing the divisor beneath the dividend, square root by writing *ka* (from the work *karana*, "irrational") before the quantity. Brahmagupta (seventh century) indicated the unknown by *yā* (from *yāvattāvat*, "so much as"). Known integers were prefixed by *rū* (from *rūpa*, "the absolute number"). Additional unknowns were indicated by the initial syllables of words for different colors. Thus a second unknown might be denoted by *kā* (from *kālaka*, "black"), and

$$8xy + \sqrt{10} - 7$$

might appear as

$$y\bar{a} \; k\bar{a} \; 8 \; bha \; ka \; 10 \; r\bar{u} \; \dot{7}.$$

Some of the Italian mathematicians of the late fifteenth and early sixteenth centuries introduced bits of syncopation into their algebra. Thus Luca Pacioli (ca. 1445—ca. 1509), in his *Summa de arithmetica* of 1494, used such abbreviations as *p* (from *piu*, "more") for plus, *m* (from *meno*, "less") for minus, *co* (from *cosa*, "thing") for the unknown *x*, *ce* (from *censo*) for x^2, *cu* (from *cuba*) for x^3, and *cece* (from *censocenso*) for x^4. Equality was sometimes indicated by *ae* (from *aequalis*).

The time was ripe for algebra to enter its symbolic stage. Thus Robert Recorde (ca. 1510-1558) gave us in 1557, in his *The Whetstone of Witte*, our modern equal sign (=). Recorde justified his adoption of a pair of equal parallel line segments for the symbol of equality "bicause noe 2 thynges can be moare equalle."

Our familiar radical sign ($\sqrt{\ }$) was introduced, because it resembles a small r, for *radix*, by Christoff Rudolff in 1525 in his book on algebra entitled *Die Coss*. The first appearance in print of our present $+$ and $-$ signs occurred in an arithmetic published in Leipzig in 1489 by Johann Widman (born ca. 1460 in Bohemia). Here the signs are not used as symbols of operation but merely to indicate excess and deficiency. Quite likely the plus sign is a contraction of the Latin word *et*, which was frequently used to indicate addition, and it may be that the minus sign is contracted from the abbreviation $\overline{m}$ for minus. Other plausible explanations have been offered. The $+$ and $-$ signs were used as symbols of algebraic operation in 1514 by the Dutch mathematician Vander Hoecke, but were probably so used earlier. In 1572, Rafael Bombelli (ca. 1526-1573) published an algebra containing improved algebraic notation. For instance, the compound expression $\sqrt{7} + \sqrt{14}$, which would have been written by Pacioli as $RV\,7\,p\,R\,14$, where RV, the *radix universalis*, indicates that the square root is to be taken of all that follows, would have been written by Bombelli as $R\ \lfloor\ 7\,p\,R\,14\ \rfloor$. Bombelli distinguished square roots and cube roots by writing $R\ q$ and $R\ c$. François Viète (1540-1603), the greatest French mathematician of the sixteenth century, did much for the development of algebraic symbolism. He used vowels to represent unknown quantities and consonants to represent known ones. It was René Descartes (1596-1650) who, in 1637, introduced our present custom of using the latter letters of the alphabet for unknowns and the early letters for knowns. Prior to Viète, it was common practice to use different letters or symbols for the various powers of a quantity; Viète used the same letter, properly qualified. Thus our x, x^2, x^3 were written by Viète as A, *A quadratum*, *A cubum*, and by later writers more briefly as A, $A\ q$, $A\ c$. It was Descartes who also introduced our present system of indices—x, x^2, x^3, and so on. Thomas Harriot (1560-1621) gave us, in his *Artis analyticae praxis*, published posthumously in 1631, our present inequality signs, $>$ and $<$. William Oughtred (1574-1660) placed great emphasis on mathematical symbols, giving over 150 of them. Of these only three have survived: the cross ($\times$) for multiplication, the four dots ($::$) used in a proportion, and our frequently used symbol for the difference between ($\sim$).

We owe to Leonard Euler (1707–1783) the conventionalization of the symbolism $f(x)$ in functional notation, e for the base of natural logarithms, Σ for the summation sign in series work, and i for the imaginary unit $\sqrt{-1}$. The symbol $n!$, called *factorial n*, was introduced in 1808 by Christian Kramp (1760–1826) of Strassbourg, who chose this symbol so as to circumvent printing difficulties incurred by the previously used symbol $\underline{n}\rfloor$. John Wallis (1616–1703) was the first to explain with any completeness the significance of zero, negative, and fractional exponents, and he introduced our present symbol (∞) for infinity. The symbol π was used by the early English mathematicians William Oughtred, Isaac Barrow (1630–1677), and David Gregory (1661–1708) to designate the circumference, or periphery, of a circle. The first to use the symbol for the ratio of the circumference to the diameter was the English writer, William Jones (1675–1749), in a publication in 1706. The symbol was not generally used in this sense, however, until Euler adopted it in 1737.

Exercises

12.1. Solve Problems 1, 2, 3, 4 of the lecture text, using high school algebra.

12.2. Thymaridas, a lesser mathematician of the fourth century B.C., gave the following rule for solving a certain set of n simultaneous linear equations connecting n unknowns. The rule became so well known that it went by the title of the *bloom of Thymaridas: If the sum of n quantities be given, and also the sum of every pair which contains a particular one of them, then this particular quantity is equal to $1/(n-2)$ of the difference between the sums of these pairs and the first given sum.*
(a) Prove the rule above.
(b) Show that Problem 5 of the lecture text is a numerical illustration of the bloom of Thymaridas.

12.3. The following problem from the *Palatine Anthology* purports to give a summary of Diophantus' personal life. "Diophantus passed one-sixth of his life in childhood, one-twelfth in youth, and one-seventh more as a bachelor. Five years after his marriage was

born a son who died five years before his father, at half his father's
[final] age." Assuming the details in the problems are correct, how
old was Diophantus when he died?

12.4. (a) How many different symbols must one memorize in
order to write numbers less than 1000 in the alphabetic Greek numeral system?

(b) In the alphabetic Greek numeral system the numbers 1000,
2000, ..., 9000 were often represented by priming the symbols for
1, 2, ..., 9. Thus 1000 might appear as α'. The number 10,000, or
myriad, was denoted by M. The multiplicative principle was used for
multiples of 10,000. Thus 20,000, 300,000, and 4,000,000 appeared
as βM, λM, and υM. Write, in alphabetic Greek, the numbers
5,780, 72,803, 450,082, 3,257,888.

(c) Make an addition table up through 10 + 10 and a multiplication table up through 10 × 10 for the alphabetic Greek numeral
system.

12.5. (a) Write $2x^4 - 21x^3 + 12x^2 - 7x + 33$ in the syncopated
notation of Diophantus.

(b) Write $3xy + 2x + 2y + \sqrt{13} - 8$ in the syncopated notation
of the Hindus.

12.6. (a) Write, in Bombelli's notation, the expression

$$\sqrt{[\sqrt[3]{(\sqrt{68} + 2)} - \sqrt[3]{(\sqrt{68} - 2)}]}.$$

(b) Bombelli indicated $\sqrt{-11}$ by *di m R q* 11. Write, in modern
notation, the following expression which occurs in Bombelli's work:

$$R c \lfloor 4 p \, di \, m \, R \, q \, 11 \rfloor p R c \lfloor 4 \, m \, di \, m \, R \, q \, 11 \rfloor.$$

12.7. Viète qualified the coefficients of a polynomial equation so
as to render the equation homogeneous, and he used our present +
and − signs, but he had no symbol for equality. Thus he would have
written

$$5BA^2 - 2CA + A^3 = D$$

as

B 5 in A quad − C plano 2 in A + A cub aequatur D solido.

Note how the coefficients C and D are qualified so as to make each term of the equation three dimensional. Write, in Viète's notation,

$$A^3 - 3BA^2 + 4CA = 2D.$$

Further Reading

CAJORI, FLORIAN, *A History of Mathematical Notations*, 2 vols. Chicago: Open Court, 1928–1929.

HEATH, T. L., *Diophantus of Alexandria*, rev. ed. New York: Cambridge University Press, 1910.

TWO EARLY COMPUTING INVENTIONS

Though the people of the world today speak and write in a great confusion of different languages, almost all of them do arithmetic with the same number symbols and utilize (pretty much) the same computing schemes.

All around the world numbers are represented, in a common, familiar way, as positional sequences of the appropriate ten *digit* symbols 0, 1, 2, 3, 4, 5, 6, 7, 8, 9. The representation neatly lends itself to the formation of succinct schemes, or patterns, called *algorithms*, for performing the operations of arithmetic. Thus there is an addition algorithm and a subtraction algorithm, a multiplication algorithm and a division algorithm, an algorithm for finding square roots and one for finding cube roots, an algorithm for finding the greatest common divisor of two given positive integers, and so on. The mastering of the basic arithmetic algorithms occupies a fair portion of a young child's school time, and he learns to perform them with the aid of two simple memorized tables called the *addition* and *multiplication tables*. This worldwide system is called, for reasons shortly to be given, the *Hindu-Arabic numeral system*.

An examination of other early numeral systems reveals these systems as, in general, less compact and/or less amenable to the formation of easy calculating algorithms. Thus, in LECTURE 12, we described the Greek alphabetic numeral system. Though this particular early numeral system does happen to represent numbers quite compactly, it does so by the employment of such a large number of symbols that one's memorization is sorely taxed. As another example, consider the Roman numeral system, familiar to most of us. Try using this system for a long multiplication or division, or, say, for the extraction of a square root. Beyond simple addition and subtraction, the system seems too awkward for the formation of easily workable algorithmic patterns.

The Hindu-Arabic numeral system is named after the Hindus, who we believe invented it, and after the Arabs, who adopted it and then transmitted it to Western Europe. The earliest preserved examples of our present number symbols are found on some stone pillars erected in India about 250 B.C. by the Maurya ruler, King Aśoka. Other early examples, if correctly interpreted, are found in India among records cut about 100 B.C. on the walls of a cave in a hill near Poona and in some inscriptions of about A.D. 200 carved in the caves at Nasik. These early specimens, however, contain no zero symbol, nor do they employ the vital idea of positional notation. Just when positional notation and a zero symbol were introduced in India is not known, but it must have been sometime before A.D. 800, for the Persian mathematician al-Khowârizmî described such a completed Hindu system in a book of A.D. 825.

How and when the new number symbols first entered Europe is not definitely settled. In all likelihood they were carried by traders and travelers on the Mediterranean coast. They are found in a tenth-century Spanish manuscript, having been introduced into Spain by the Arabs, who invaded the peninsula in A.D. 711 and remained there for several hundred years. The system, with its early algorithms, was more widely disseminated by a twelfth-century Latin translation of al-Khowârizmî's treatise, and by subsequent European works on the subject, a particularly influential one of which will be discussed in a later lecture.

Why, one might wonder, didn't the ancients devise more workable numeral systems—systems more amenable to simple computing patterns? In other words, why wasn't a system similar to the Hindu-Arabic system devised much earlier, along with algorithms for adding, subtracting, multiplying, dividing, and extracting square and cube roots? There is a very practical reason that accounts for the tardiness of such an invention. Without a plentiful, convenient, and cheap supply of some suitable writing material, the seeking of such a numeral system would hardly even occur to anyone. It must be remembered that our common machine-made pulp paper is only somewhat over a hundred years old. The older rag paper was made by hand, was consequently expensive and rather scarce, and, even at that, was not introduced into Europe until the twelfth century, although it is likely that the Chinese knew how to make it a thousand years before. Earlier writing materials—like stone and chisel, clay tablets, sand trays, papyrus, parchment, and vellum—were either too inconvenient or too hard to come by to be used for mere scratch work; and much scratch work, in the form of trial and error and experimentation, is needed for evolving good computing algorithms. Small white boards covered with a thin sprinkling of red flour, wax-coated boards, and slates were generally of later use, and, along with sharp styluses, were quite suitable for the development of early arithmetic algorithms. Because of limited area and ease of erasure, these early algorithms were devised with conservation of space in mind, and digits in an algorithm were effaced as soon as they had served their purpose. The result was that a computation could be checked only by repeating the whole process over again. It is not often realized that most of our present-day computing patterns used in conjunction with the Hindu-Arabic numeral system did not reach final form until the thirteenth century.

How, then, did early peoples perform their arithmetical calculations? The way around the difficulties mentioned above was by the simple but remarkable invention of the abacus.

The abacus is a very old (just how old is not known) device that assumed different forms in different parts of the ancient and medieval world, and, next to man's fingers, constitutes the oldest calculating device of mankind. With it one can add, subtract, multiply, divide, and extract square and cube roots. In the hands of a dexterous and practiced manipulator, the instrument can perform these arithmetical operations with astonishing speed and accuracy.

The most primitive form of the abacus seems to have been a marked sand tray, whence comes the generic name (from *abax*, Greek for "sand tray"). Forms of the abacus were used by the ancient Egyptians, Greeks, Romans, Hindus, and Far Eastern peoples. A more sophisticated form of the abacus was a board with small pebbles (*calculi*) sliding in grooves; another was a wooden frame with beads sliding on wires or thin bamboo rods.

Let us describe a rudimentary form of abacus and illustrate its use in the addition of a pair of Roman numbers. Draw four vertical parallel lines and label them from left to right by M, C, X, and I, and obtain a collection of convenient counters, like checkers or pennies. A counter will represent 1, 10, 100, or 1000 according as it is placed on the I, X, C, or M line. To reduce the number of counters which may subsequently appear on a line, we agree to replace any five counters on a line by one counter in the space just to the left of that line. These spaces, then, may be labeled from left to right by D, L, and V. Any number less than 10,000 may now be represented on our frame of lines by placing not more than four counters on any line and not more than one counter in the space between two lines.

Let us now add

MDCCLXIX   and   MXXXVII.

Represent the first of the two numbers by counters on the frame, as illustrated on the left in Figure 44. We now proceed to add the second number, working from right to left. To add the VII, put another counter between the X and I lines and two more counters on the I line. The I line now has six counters on it. We remove five of them and put instead another counter between the X and I lines. Of the three counters now between the X and I lines, we "carry over" two of them as a single counter on the X line. We now add the XXX by putting three more counters on the X line. Since we now have a total of five counters on the X line, they are replaced by a single counter between the C and X lines, and the two counters now found there are "carried over" as a single counter on the C line. We finally add the M by putting another counter on the M line. The final appearance of our frame is illustrated on the right in Figure 44 and the sum can be read off as MMDCCCVI. We have obtained the sum of the two numbers by simple mechanical operations and without requiring any scratch paper or recourse to memorization of any addition tables.

Subtraction is similarly carried out, except that now, instead of "carrying over" to the left, we may find it necessary to "borrow" from the left. Consider, for example, the problem of subtracting

MDCCLXIX   from   MMDCCCVI.

Represent the second of these two numbers by counters on the abacus frame, as illustrated on the left in Figure 45. We proceed to subtract the first number, working from right to left. Since we can-

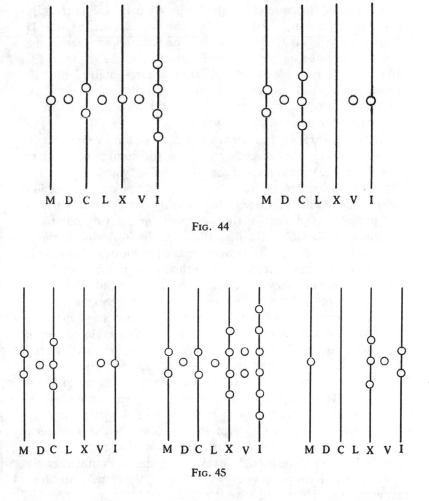

FIG. 44

FIG. 45

not subtract the IX from VI, we must first do some "borrowing" from the left. Take one of the counters from the C line and replace it by one in the L space and five on the X line, and then replace one of the counters on the X line by one in the V space and five on the I line. The frame now looks as pictured in the center in Figure 45. We can now subtract the IX, which is VIIII, leaving only one counter in the V space and two counters on the I line. We may now subtract, in turn, one X counter, the L counter, the two C counters, the D counter, and one of the M counters. The final appearance of the frame is illustrated on the right in Figure 45 and the required difference can be read off as MXXXVII. We have obtained the difference of the two numbers by simple mechanical operations, without the use of any scratch paper and without recourse to memorization of any arithmetic tables.

The Hindu-Arabic positional numeral system represents a number very simply by recording in order the number of counters belonging to the various lines of the abacus. The symbol 0 stands for a line with no counters on it. Our present addition and subtraction patterns, along with the concepts of "carrying over" and of "borrowing" probably originated in the processes for carrying out these operations on the abacus. Since, with the Hindu-Arabic numeral system, we are working with symbols instead of the actual counters, it becomes necessary either to commit the simple number combinations to memory or to have recourse to an elementary addition table.

Europeans of the Middle Ages who advocated the use of Roman numerals along with the abacus for calculation were called *abacists*, and those who advocated the use of the Hindu-Arabic numerals along with appropriate algorithms for calculation were called *algorists*. The four hundred years from 1100 to 1500 witnessed the long, and sometimes bitter, battle between the abacists and the algorists. By 1500 our present rules in computing won supremacy. In another hundred years the abacists were almost forgotten, and by the eighteenth century no trace of an abacus was found in western Europe. Its reappearance, as a curiosity, was due to the French geometer Poncelet, who brought back a specimen to France after his release as a Russian prisoner of war following the Napoleonic Russian campaign. Though now most of the world is converted to the Hindu-Arabic numeral system and its algorithms, there are parts of the Eastern world that still use the abacus. Thus there is the *suan pan* of China, the *soroban* of Japan, the *tschoty* of Russia, and the similar reckoning frame of some of the Arabian countries.

There was considerable variation found in the Hindu-Arabic number symbols until they became stabilized by the development of printing. Our word *zero* comes from the Latinized form *zephirium* of the Arabic *sifr*, which in turn is a translation of the Hindu *sunya*, meaning "void" or "empty." The Arabic *sifr* was introduced into Germany, in the thirteenth century by Nemorarius, as *cifra*, from which we have obtained our present word *cipher*.

It is easy today to criticize the abacists as having been overly conservative and fearful of change. But it was only natural to experience a feeling of insecurity with the new numerals; they were unfamiliar and, for a long time, were not standardized; the zero numeral was particularly confusing; there was much that had to be patiently learned before one could properly use the new symbols. But the abacists had a much stronger and much more cogent objection to the new numerals: the new numerals too easily lent themselves to fraud. One could easily turn a 0 into a 6 or a 9, or turn a 1 into a 4, a 6, a 7, or a 9. Other numeral forms could also be tampered with, and the insertion of numerals between or after some already recorded was often quite possible. It was for reasons of such possible falsification that the City Council of Florence, for example, in 1299 issued its ordinance prohibiting under fine the use of the new numerals in financial procedures.

In this lecture we have considered two remarkable early inventions that have greatly aided mankind in his task of performing calculations: the abacus and the Hindu-Arabic numeral system. Each invention is of uncertain date of origin—the Hindu-Arabic numeral system certainly appeared sometime before A.D. 800, and the abacus at a much earlier time. There is no question, however, that the appearance of each of these two great historically related inventions marks a GREAT MOMENT IN MATHEMATICS.

Although most of the world has now gone the way of the algorists, schoolteachers of today in the United States, France, and other countries, still find the abacus useful when teaching young pupils the idea of place value in our Hindu-Arabic numeral system; and we still refer to the table where the clerk of a store figures the customers' bills as "the counter." We have already seen that our concepts of "carrying over" in addition and "borrowing" in subtraction in all likelihood originated in the processes for performing these operations on an abacus. Our word "calculate" derives from the Latin word *calculus*, meaning "pebble," and pebbles served as the counters on the Roman abacus. Many games of today played with collections of like flat round pieces owe their origin to the time when abacus counters were common household objects.

We conclude with an interesting remark on our own mixed culture: Our language is Germanic, our writing is Roman, and our numerals are Indian.

Exercises

13.1. Using a primitive abacus find the sum and difference of

MDCCLXXXIX   and   MMDCLXXVIII.

13.2. The Hindu-Arabic numeral system is an example of a *positional numeral system* with *base* 10. One can have a positional numeral system with any integral base $b > 1$. After the base b has been selected, one must adopt b basic symbols for 0, 1, 2, ..., $b - 1$; these basic symbols are called the *digits* of the system. If $b \leq 10$, we may use our familiar Hindu-Arabic digit symbols; if $b > 10$, we may use our familiar Hindu-Arabic digit symbols augmented by an appropriate number of extra symbols.

Show that any (whole) number N can be expressed uniquely in the form

$$N = a_n b^n + a_{n-1} b^{n-1} + \cdots + a_2 b^2 + a_1 b + a_0,$$

where $0 \leq a_i < b$, $i = 0, 1, ..., n$.

The number N above is represented positionally with base b by the sequence of digit symbols

$$a_n a_{n-1} \cdots a_2 a_1 a_0.$$

It follows that any one of these digit symbols represents a multiple of some power of the base, the power depending upon the position in which the digit occurs. Thus, for example, we may consider 3012 as a number expressed positionally with base 4 and with digit symbols 0, 1, 2, 3. To make clear that the number is considered as expressed with base 4 we shall write it as $(3012)_4$. When no subscript is written it will be understood that the number is expressed with the ordinary base 10.

13.3. (a) Construct addition and multiplication tables for base 7.

(b) Now add $(3406)_7$ and $(251)_7$.

(c) Multiply $(3406)_7$ and $(251)_7$.

13.4. (a) Construct addition and multiplication tables for base 12, using the symbols t and e for the digits ten and eleven.

(b) Now add $(3t04e)_{12}$ and $(51tt)_{12}$.

(c) Multiply $(3t04e)_{12}$ and $(51tt)_{12}$.

13.5. (a) Express $(3012)_5$ in base 8.

(b) For what base is $3 \times 3 = 10$? For what base is $3 \times 3 = 11$? For what base is $3 \times 3 = 12$?

(c) Can 27 represent an even number in some scale? Can 37? Can 72 represent an odd number in some scale? Can 82?

(d) Find b such that $79 = (142)_b$.

(e) Find b such that $72 = (2200)_b$.

13.6. (a) A three-digit number in the scale of 7 has its digits reversed when expressed in the scale of 9. Find the three digits.

(b) What is the smallest base for which 301 represents a square integer?

(c) If $b > 2$, show that $(121)_b$ is a square integer.

(d) If $b > 4$, show that $(40001)_b$ is divisible by $(221)_b$.

13.7. The positional numeral system with base 2 (called the *binary system*) has applications in various branches of mathematics. Also, there are many games and puzzles, like the well-known game of *Nim* and the puzzle of the *Chinese rings*, having solutions that depend on this system. Following are two easy puzzles of this sort.

(a) Show how to weigh, on a simple equal-arm balance, any weight w of a whole number of pounds using a set of weights of 1 pound, 2 pounds, 2^2 pounds, 2^3 pounds, and so forth, there being only one weight of each kind.

(b) Consider the four cards illustrated in Figure 46 containing numbers from 1 through 15. On the first card are all those numbers whose last digit in the binary system is 1; the second contains all those numbers whose second digit from the end is 1; the third contains all those whose third digit from the end is 1; the fourth contains all those whose fourth digit from the end is 1. Now someone is asked to think of a number N from 1 through 15 and to tell on which cards N can be found. It is then easy to announce the number N by

1	9
3	11
5	13
7	15

2	10
3	11
6	14
7	15

4	12
5	13
6	14
7	15

8	12
9	13
10	14
11	15

Fig. 46

merely adding the top left numbers on the cards where it appears. Now make a similar set of six cards for detecting any number from 1 through 63. If the numbers are written on cards weighing 1, 2, 4, ... units, then an automaton in the form of a postal scale can express the number N.

13.8. Many simple number tricks, where one is to "guess a selected number," have explanations depending on our own positional scale. Expose the following tricks of this kind.

(a) Someone is asked to think of a two-digit number. He is then requested to multiply the tens digit by 5, add 7, double, add the units digit of the original number, and announce the final result. From this result the conjurer secretly subtracts 14 and obtains the original number.

(b) Someone is asked to think of a three-digit number. He is then requested to multiply the hundreds digit by 2, add 3, multiply by 5, add 7, add the tens digit, multiply by 2, add 3, multiply by 5, add the units digit, and announce the result. From this result the conjurer secretly subtracts 235 and obtains the original number.

(c) Someone is asked to think of a three-digit number whose first and third digits are different. He is then requested to find the difference between this number and that obtained by reversing the three digits, by subtracting the smaller from the larger. Upon disclosing only the last digit in this difference, the conjurer announces the entire difference. How does the trickster do this?

13.9. The arithmetics of the fifteenth and sixteenth centuries contain descriptions of algorithms for the fundamental operations. Of the many schemes devised for performing a long multiplication, the so-called *gelosia*, or *grating*, method was perhaps the most popular. The method, which is illustrated in Figure 47 by the multiplication of 9876 and 6789 to yield 67,048,164, is very old. It was probably first developed in India, for it appears in a commentary on the *Lilāvati* of Bhāskara (1114–ca. 1185) and in other Hindu works. From India it made its way into Chinese, Arabian, and Persian works. It was long a favorite method among the Arabs, from whom it passed over to the western Europeans. Because of its simplicity to apply, it could well be that the method might still be in use but for the difficulty of printing, or even drawing, the needed net of lines. The pattern resembles the grating, or lattice, used in some windows. These were known as "gelosia," eventually becoming "jalousie" (meaning "blind" in French).

Find the product of 80,342 and 7,318 by the gelosia method.

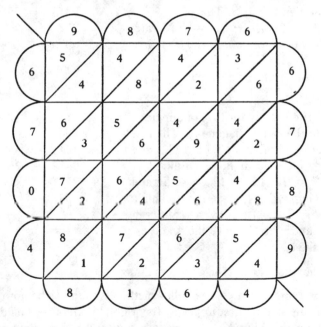

FIG. 47

13.10. By far the most common algorithm for long division in use before 1600 was the so-called *galley*, or *scratch*, method, which in all likelihood was of Hindu origin. To illustrate the method, consider the following steps in the division of 9413 by 37.

1. Write the divisor, 37, below the dividend as shown. Obtain the first quotient digit, 2, in the usual manner, and write it to the right of the dividend.

9413 | 2
37

2. Think: $2 \times 3 = 6, 9 - 6 = 3$. Scratch 9 and 3 and write 3 above the 9. Think: $2 \times 7 = 14, 34 - 14 = 20$. Scratch 7, 3, 4 and write 2 above the 3 and 0 above the 4.

2
3̶0
9̶4̶1̶3 | 2
3̶7̶

3. Write the divisor, 37, one place to the right, diagonally. The resultant dividend after Step 2 is 2013. Obtain the next quotient digit, 5. Think: $5 \times 3 = 15, 20 - 15 = 5$. Scratch 3, 2, 0 and write 5 above the 0. Think: $5 \times 7 = 35, 51 - 35 = 16$. Scratch 7, 5, 1 and write 1 above the 5 and 6 above the 1.

1
2̶5̶
3̶0̶6
9̶4̶1̶3 | 25
3̶7̶7
3̶

4. Write the divisor, 37, one more place to the right, diagonally. The resultant dividend after Step 3 is 163. Obtain the next quotient digit, 4. Think: $4 \times 3 = 12, 16 - 12 = 4$. Scratch 3, 1, 6 and write 4 above the 6. Think: $4 \times 7 = 28, 43 - 28 = 15$. Scratch 7, 4, 3 and write 1 above the 4 and 5 above the 3.

1̶1
2̶3̶4̶
3̶0̶6̶5
9̶4̶1̶3 | 254 —|1|
3̶7̶7̶7 |5|
3̶3̶

5. The quotient is 254, with remainder 15.

After a little practice, the galley method is found to be not nearly as difficult as it at first appears. Its original popularity was due to the ease with which it could be used on a sand tray or slate, where the scratching was a simple erasing followed by a possible replacement. Later, when paper and ink were used, erasures were not easily effected, and undesired digits were merely crossed off and, when wanted, new ones written above the old ones, as we have done in the illustration above. This procedure led to the name "scratch" method. The name "galley" method arose from the fancied resemblance of the outline of the finished problem to a boat. The resemblance follows either by viewing the work from the bottom of the page, when the quotient appears as a bowsprit, or by viewing the work from the left side of the page, when the quotient appears as a mast. In this second viewpoint, the remainder was frequently written (as indicated above) like a flag at the top of the mast.

Divide 65,284 by 594, using the galley method. (This problem, solved in this way, appears in the *Treviso Arithmetic* of 1478.)

Further Reading

HILL, G. F., *The Development of Arabic Numerals in Europe.* New York: Oxford University Press, 1915.

KARPINSKI, LOUIS CHARLES, *The History of Arithmetic.* New York: Russell & Russell, 1965.

PULLAN, J. M., *The History of the Abacus.* New York: Praeger, 1968.

THE POET-MATHEMATICIAN OF KHORASAN

In the second half of the eleventh century, three Persian youths, each a capable scholar, studied together as pupils of one of the greatest wise men of Khorasan, the Imam Mowaffak of Naishapur. The three youths—Nizam ul Mulk, Hasan Ben Sabbah, and Omar Khayyam—became close friends. Since it was the belief that a pupil of the Imam stood great chance of attaining fortune, Hasan one day proposed to his friends that the three of them take a vow to the effect that, to whomever of them fortune should fall, he would share it equally with the others and reserve no preeminence for himself. As the years went by, Nizam proved to be the fortunate one, for he became Vizier to the Sultan Alp Arslan. In time his school friends sought him out and claimed a share of his good fortune according to the school-day vow.

Hasan demanded a governmental post, which was granted by the Sultan at the Vizier's request. But, being selfish and ungrateful, he endeavored to supplant his friend Nizam and was finally disgraced and banished. Omar desired neither title nor office, but simply begged to be permitted to live in the shadow of the Vizier's fortune, where he might promulgate science and mathematics and pray for his friend's long life and prosperity. Impressed by his former schoolmate's modesty and sincerity, the Vizier granted Omar a yearly pension.

After many misadventures and wandering, Hasan became the head of a party of fanatics who, in 1090, seized the castle of Alamut in the mountainous area south of the Caspian Sea. Using the castle as a fortress and center for raids upon passing caravans, Hasan and his band spread terror through the Mohammedan world. Hasan became known as "the old man of the mountain," and it is thought that our present word "assassin" derives either from the leader's name *Hasan* or from the *hashish* opiate with which the band maddened themselves for their murderous assaults. Among the countless victims of the assassins was the old school friend, Nizam ul Mulk. In contrast to Hasan's turbulent and destructive life, Omar's was tranquil and constructive. He lived peacefully and contributed noteworthily to both the literary and the scientific culture of his age. So, of the three students, one turned out to be a fine administrator and benefactor, one a miserable renegade and murderer, and one a devoted scholar and creator. Our present lecture will be devoted to a remarkable mathematical accomplishment of the scholar—an accomplishment that ranks as one of the GREAT MOMENTS IN MATHEMATICS. We first dispose of some preliminary groundwork.

By a *real polynomial equation* in one unknown x is meant any equation of the form

$$a_0x^n + a_1x^{n-1} + \cdots + a_{n-1}x + a_n = 0,$$

where n is a positive integer and $a_0, a_1, \ldots, a_n$ are real numbers with $a_0 \neq 0$. A value of x that satisfies the equation is called a *root* of the equation. One of the principal tasks of early algebra was to try to find general methods for obtaining the roots of such equations; this was known as *solving* the equations. Since the only numbers known in early days were the positive real numbers, to solve an equation meant, for many hundreds of years, to find the positive real roots, if any, possessed by the equation. The equations are said to be *linear, quadratic, cubic, quartic, quintic,* ... according as $n = 1, 2, 3, 4, 5, \ldots$.

Now the linear case presents no difficulty and can easily be solved either geometrically or algebraically. If a linear polynomial equation in one unknown has a positive root, the equation can always be written in the form

$$ax = b,$$

where a and b are both positive numbers. Algebraically, it follows that $x = b/a$. Geometrically, x is the fourth proportional to the three segments having lengths a, b, and 1. That is, $a:b = 1:x$, and x can be found (actually with straightedge and compasses) by the simple construction indicated in Figure 48, where *COD* is any conve-

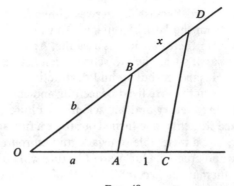

FIG. 48

nient angle, $OA = a$, $OB = b$, $AC = 1$, and CD is drawn parallel to AB.

It is interesting that the ancient Egyptians solved linear equations by a method later known in Europe as the *rule of false position*. Thus, to solve

$$x + x/7 = 24,$$

assume any convenient value for x, say $x = 7$. Then $x + x/7 = 8$, instead of 24. Since 8 must be multiplied by 3 to give the required 24, the correct x must be $3(7) = 21$. It is fascinating to contemplate that, from a sheer guess, the correct answer has been found.

Though the quadratic case is, of course, more complicated than the linear case, mathematicians of early times solved this case, too, both geometrically and algebraically. The algebraic solution, either by the method of completing the square or by substitution in the general quadratic formula, should be familiar to anyone who has had elementary high school algebra. The ancient Babylonians of approximately 2000 B.C. knew the equivalents of both of these algebraic methods. In Exercise 8.10 (of LECTURE 8), the student can find what is essentially the Greek geometrical solution of quadratic equations.

The cubic case is a considerably more difficult matter, and though, with the aid of a table giving values of $n^3 + n^2$ for given values of n, the ancient Babylonians solved certain special cubic equations, and Archimedes, in a fragment preserved for us by the commentator Eutocius, discussed conditions under which a cubic may have a real and positive root, a general algebraic solution of cubic equations had to await work done by Italian mathematicians in the sixteenth century. A *geometric* solution of cubic equations, however, was found almost a half-millennium earlier, in the eleventh century, by the Persian poet-mathematician Omar Khayyam. This is the GREAT MOMENT IN MATHEMATICS alluded to above, and we now consider it, after first laying a little historical background. The algebraic solution of cubic equations will be considered in LECTURE 16.

The period from the fall of the Roman Empire in the middle of the fifth century to the eleventh century is known as Europe's Dark Ages, for during this period learning and civilization in Western Europe sank to a very low ebb. This same period, however, marks the spectacular rise of the Arabian Empire. Within a decade following Mohammed's flight from Mecca to Medina in A.D. 623, the scattered and divided Bedouin tribes of the Arabian peninsula were united by a strong religious fervor into a powerful nation. Within a century, force of arms under the green and gold banner of Islam had extended the rule of the Moslem star and crescent over a territory that reached from India, through Persia, Mesopotamia, and northern Africa, clear into Spain.

Of great importance for the preservation of much of world culture was the assiduous manner in which the Arabs seized upon Greek and Hindu erudition. Numerous Hindu and Greek works in medicine, astronomy, and mathematics were sedulously translated into the Arabic tongue and thus were saved until later European scholars were able to retranslate them into Latin and other languages. But for the industrious work of the Arabian scholars much of Greek and Hindu science would have been irretrievably lost over the long period of the Dark Ages. In addition to serving admirably as preservers of much of the world's intellectual possessions, the Arabs made some advances of their own, and highly original among these advances was the work done by Omar Khayyam in connection with the geometrical solution of cubic equations.

Omar Khayyam (ca. 1044–ca. 1123) was a Persian poet, astronomer, and mathematician, born and educated in Naishapur of

Khorasan (now Neyshābūr in northeastern Iran). He is widely and affectionately known in the Western world through the beautiful and felicitous translation by Edward Fitzgerald of his exquisite verses known as *The Rubaiyat*. To the scientific world, Omar Khayyam is also noted for his remarkably accurate calendar reform, his critical treatment of Euclid's parallel postulate showing him to be a forerunner of Saccheri's ideas that finally led to the creation of non-Euclidean geometry, and, particularly, for his original contribution to Arabian algebra wherein he managed to solve geometrically, so far as positive roots are concerned, every type of cubic equation.

Let us illustrate Omar Khayyam's procedure for the special type of cubic

$$x^3 + b^2x + a^3 = cx^2,$$

where a, b, c, x are thought of as lengths of line segments. Omar stated this type of cubic rhetorically as "a cube, some sides, and some numbers are equal to some squares." Stated geometrically, the problem of solving the cubic equation is this: Given a unit segment and the line segments a, b, c, construct a line segment x such that the above relation among a, b, c, x will hold. The object is to construct x using only straightedge and compasses as far as possible. A solution using only straightedge and compasses is in general impossible, and at some point of the construction we must be permitted to draw a certain uniquely defined conic section.

A basic construction, used several times in the solution of the cubic, is that of finding the fourth proportional to three given line segments. This is an old problem whose solution was known to the ancient Greeks. Suppose u, v, w are three given line segments and we desire a line segment x such that $u:v = w:x$. Figure 49, which is similar to Figure 48, will recall how, with straightedge and compasses, one may construct the desired segment x.

We now follow Omar's geometrical solution of the cubic equation

$$x^3 + b^2x + a^3 = cx^2.$$

First of all, by the basic construction, find line segment z such that $b:a = a:z$. Then, again by the basic construction, find line segment m such that $b:z = a:m$. We easily find that $m = a^3/b^2$. Now, in

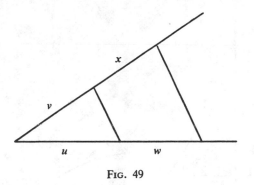

FIG. 49

Figure 50, draw $AB = m = a^3/b^2$ and $BC = c$. Draw a semicircle on AC as diameter and let the perpendicular to AC at B cut it in D. On BD mark off $BE = b$ and through E draw EF parallel to AC. By the basic construction, find G on BC such that $ED:BE = AB:BG$ and complete the rectangle $DBGH$. Through H draw the rectangular hyperbola having EF and ED for asymptotes (that is, the hyperbola through H whose equation with respect to EF and ED as x and y axes is of the form $xy =$ a constant). Let the hyperbola cut the semicircle in J, and let the parallel to DE through J cut EF in K and BC in L. Let GH cut EF in M. Now:

   1. Since J and H are on the hyperbola, $(EK)(KJ) = (EM)(MH)$.
   2. Since $ED:BE = AB:BG$, we have $(BG)(ED) = (BE)(AB)$.
   3. Therefore, from 1 and 2, $(EK)(KJ) = (EM)(MH) = (BG)(ED) = (BE)(AB)$.
   4. Now $(BL)(LJ) = (EK)(BE + KJ) = (EK)(BE) + (EK)(KJ) = (EK)(BE) + (AB)(BE)$ (by 3) $= (BE)(EK + AB) = (BE)(AL)$, whence $(BL)^2(LJ)^2 = (BE)^2(AL)^2$.
   5. But, from elementary geometry, $(LJ)^2 = (AL)(LC)$.
   6. Therefore, from 4 and 5, $(BE)^2(AL) = (BL)^2(LC)$, or $(BE)^2(BL + AB) = (BL)^2(BC - BL)$.
   7. Setting $BE = b$, $AB = a^3/b^2$, $BC = c$ in 6, we obtain $b^2(BL + a^3/b^2) = (BL)^2(c - BL)$.
   8. Expanding the last equation in 7, and arranging terms, we find $(BL)^3 + b^2(BL) + a^3 = c(BL)^2$, and it follows that $BL = x$, a root of the given cubic equation.

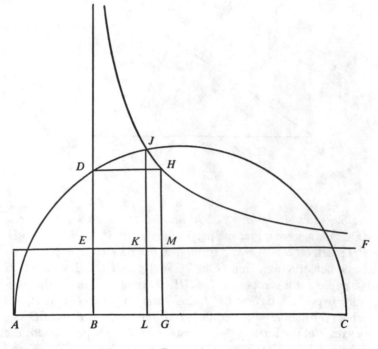

Fig. 50

It must be admitted that Omar's method is ingenious, and a high school teacher certainly should be able to interest some of his students in it. A point-by-point construction of the hyperbola (utilizing the basic construction) is easy, for if N is any point on EF and if the perpendicular to EF at N cuts the hyperbola in P, then $(EM)(MH) = (EN)(NP)$, whence $EN:EM = MH:NP$, and NP is the fourth proportional to the three given segments EN, EM, MH. In this way a number of points on the hyperbola can be plotted and then the hyperbola sketched in by drawing a smooth curve through the plotted points. The student might be given the numerical cubic $x^3 + 2x + 8 = 5x^2$. Here $a = 2$, $b = \sqrt{2}$, $c = 5$. The three roots of the cubic are 2, 4, -1. The student should be able to find the two positive roots by Omar's method; perhaps he can extend the method slightly to find the negative root. Here is the start of an excellent "junior" research project. The student can find Omar's geometrical approach to other types of cubics in J. L. Coolidge, *The Mathematics of Great Amateurs* (Oxford, 1950), Chapter II (Omar Khayyam), pp. 19-29.

Omar Khayyam died in Naishapur about 1123. A pupil of his, one Khwajah Nizami of Samarkand, has related that he used to converse with his teacher Omar in a garden, and that Omar once said that his tomb would be located in a spot where the north wind would scatter rose petals over it. Some years later, after the death of his teacher, the former pupil chanced to visit Naishapur and he searched out the master's grave. It was just outside a garden. Boughs of fruit trees hanging over the garden wall had dropped so many blossoms on the grave that the tombstone was completely hidden.

When Edward Fitzgerald, the sympathetic Irish translator who made Omar Khayyam so famous in modern times, died in 1883, he was buried in a little English churchyard at Boulge, Suffolk. In 1884, William Simpson, a traveling artist of the *Illustrated London News*, visited Naishapur and found the not-quite-neglected tomb of Omar. Along the edge of the platform in front of the tomb he found some rose trees and he plucked from these a few of the hips still hanging on the bushes. These seeds, when they arrived in England, were handed over to Mr. Baker of the Kew Botanical Gardens, who planted them and successfully grew some rose trees from them. On October 7, 1893, one of these rose trees was transplanted to Fitzgerald's graveside.

> Look to the blowing Rose about us—"Lo,
> Laughing," she says, "into the world I blow,
>   At once the silken tassel of my Purse
> Tear, and its Treasure on the Garden throw."

Exercises

14.1. Solve, by the rule of false position, the following problem found in the Rhind papyrus (ca. 1650 B.C.): "A quantity, its $2/3$, its $1/2$, and its $1/7$, added together, become 33. What is the quantity?"

14.2. We wish to find the length of side BC of the quadrilateral pictured in Figure 51. Use the rule of false position by taking BC to be any convenient length, say 1 unit, and on this basis calculate, by the law of sines, BA and BD, and then, by the law of cosines, AD; etc. (This is a highly practical procedure for solving the problem and is essentially the method a good surveyor might use.)

14.3. In the study of geometrical constructions there is a counterpart of the rule of false position, known as the *method of similitude*. The method lies in constructing a figure similar to the one desired, and then, by the use of proportion, "blowing it up" to proper size. Suppose, for example, we wish to inscribe a square in a given triangle ABC so that one side of the square lies along the base BC of the triangle (see Figure 52). First draw a square $D'E'F'G'$ of any convenient size, as indicated. If F' falls on AC, the problem is solved. Otherwise we have solved the problem for a triangle $A'BC'$ similar to triangle ABC and having B as a center of similitude. It follows that line BF' cuts AC in the vertex F of the sought square inscribed in triangle ABC.

Construct, by the method of similitude, a line segment DE, where D is on side AB and E on side AC of a given triangle ABC, so that $BD = DE = EC$.

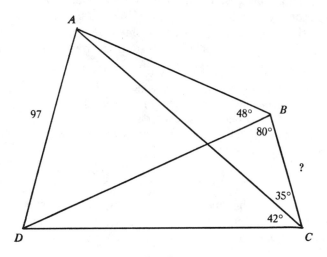

FIG. 51

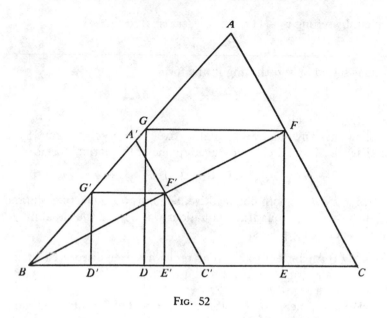

Fɪɢ. 52

14.4. (a) An ancient Babylonian problem asks for the side of a square if the area of the square diminished by the side of the square is 870. The solution of the problem is described as follows: "Take half of 1, which is ½; multiply ½ by ½, which is ¼; add the ¼ to 870, to obtain 870¼. This last is the square of 29½. Now add ½ to 29½; the result is 30, which is the side of the square." Show that this Babylonian solution is exactly equivalent to solving the quadratic equation

$$x^2 - px = q$$

by substituting in the formula

$$x = \sqrt{(p/2)^2 + q} + p/2.$$

(b) Another ancient Babylonian text solves the quadratic equation

$$11x^2 + 7x = 6\tfrac{1}{4}$$

by first multiplying through by 11 to obtain

$$(11x)^2 + 7(11)x = 68\tfrac{3}{4},$$

which, by setting $y = 11x$, has the "normal form"

$$y^2 + py = q.$$

This is solved by substituting in the formula

$$y = \sqrt{(p/2)^2 + q} - p/2.$$

Finally, $x = y/11$.

Show that any quadratic equation $ax^2 + bx + c = 0$ can, by a similar transformation, be reduced to one of the normal forms

$$y^2 + py = q, \qquad y^2 = py + q, \qquad y^2 + q = py,$$

where p and q are both nonnegative. So far as we know, the solution of such three-term quadratic equations was beyond the capabilities of the ancient Egyptians.

14.5. (a) An ancient Babylonian tablet has been discovered which gives the values of $n^3 + n^2$ for $n = 1$ through 30. Make such a table for $n = 1$ through 10.

(b) Find, by means of the table above, a root of the cubic equation $x^3 + 2x^2 - 3136 = 0$.

(c) A Babylonian problem of about 1800 B.C. seems to call for the solution of the simultaneous system

$$xyz + xy = {}^7\!/\!{}_6, \qquad y = 2x/3, \qquad z = 12x.$$

Solve this system using the table of part (a).

(d) Otto Neugebauer (b. 1899) believes that the ancient Babylonians were quite capable of reducing the general cubic equation to the "normal form" $y^3 + y^2 = c$, although there is as yet no evidence that they actually accomplished this. Show how such a reduction might be made.

14.6. Given segments of lengths a, b, c, construct with straightedge and compasses a segment of length $m = a^3/bc$.

14.7. (a) Find geometrically, by Omar Khayyam's method, the positive roots of the cubic equation $x^3 + 2x + 8 = 5x^2$.

(b) Extend the method slightly to find the negative root.

14.8. (a) Show the incomplete cubic equation $ax^3 + bx + c = 0$ can be solved geometrically for its real roots on a rectangular cartesian coordinate framework on which the cubic curve $y = x^3$ has already been drawn by merely drawing the line $ay + bx + c = 0$.

(b) Solve, by the method of part (a), the cubic equation $x^3 + 6x - 15 = 0$.

(c) Solve the cubic equation $4x^3 - 39x + 35 = 0$ geometrically.

(d) Show that any complete cubic equation $ax^3 + bx^2 + cx + d = 0$ can be reduced to the incomplete form in the variable z by the substitution $x = z - b/3$.

(e) Now solve the cubic equation $x^3 + 9x^2 + 20x + 12 = 0$ geometrically.

It is interesting that any complex imaginary roots possessed by either an incomplete or a complete cubic equation can also be found geometrically. See, e.g., Arthur Schultze, *Graphic Algebra* (New York: Macmillan Company, 1922), Sections 58, 59, 65.

Further Reading

COOLIDGE, J. L., *The Mathematics of Great Amateurs.* New York: Oxford University Press, 1949.

THE BLOCKHEAD

The dissemination and popularization of the Hindu-Arabic numeral system in western Europe was largely accomplished by the publication of certain books extolling and advocating the new numerals.

The earliest Arabic arithmetic known to us is that of al-Khowârizmî (ca. 825), which was followed by a number of other Arabic arithmetics by later authors. These arithmetics contained rules, fashioned after the Hindu patterns, for computing with the Hindu numerals. They also gave the process known as *casting out 9's*, used for checking arithmetical calculations, and the rules of *false position* and *double false position*, by which certain algebra problems can be solved nonalgebraically. Square and cube roots, fractions, and the *rule of three* were also frequently explained.

Al-Khowârizmî's book on the use of the Hindu numerals introduced a word into the common vocabulary of mathematics. This book is not extant in the original, but in 1857 a Latin translation was found which begins, "Spoken has Algoritmi, ..." Here the name *al-Khowârizmî* has become *Algoritmi,* from which, in turn, was derived our present word "algorithm," meaning a formal procedure of calculating in some particular way.

It is interesting that al-Khowârizmî is responsible for another word now part of the common vocabulary of mathematics—the word "algebra." The word comes from the title, *Hisâb al-jabr w'al-muqâbalah,* of al-Khowârizmî's treatise on the subject. The title may be literally translated as "Science of the reunion and the opposition," or more freely as "Science of transposition and cancellation."* The text, which is extant, became known in Europe through Latin translations, and made the word *al-jabr*, or *algebra*, for a long time synonymous with the science of equations.

The Arabic word *al-jabr*, used in a nonmathematical sense, found its way into Europe through the Moors of Spain. There an *algebrista* was a bone-setter (reuniter of broken bones), and it was usual for a barber of the time to call himself an *algebrista*, for bone-setting and bloodletting were sidelines of the medieval barber. This old practice of barbers is reflected in the familiar red and white striped barber poles of today.

Let us, however, return to works that played an influential role in spreading knowledge and use of the Hindu-Arabic numeral system. By far the most influential was a book called the *Liber abaci* that appeared in Italy in the year 1202. The appearance of this work was a GREAT MOMENT IN MATHEMATICS, and it is largely about this work that the present lecture will be concerned.

The author of the *Liber abaci* was Leonardo Fibonacci ("Leonardo, son of Bonaccio," 1175–1250?), the most skilled mathematician of the Middle Ages. Also known as Leonardo of Pisa (or Leonardo Pisano), Fibonacci was born in the commercial town of Pisa, where his father was connected with the mercantile business. Many of the large Italian business firms in those days maintained warehouses in various parts of the Mediterranean world. It was in this way, when his father was serving as a customs manager, that young Leonardo was brought to Bougie, on the north coast of Africa. The father's occupation early roused in the boy an interest in arithmetic, and subsequent visits to Arabian ports and extended trips to Egypt, Sicily, Greece, and Syria brought him in contact with Eastern and Arabian mathematical practices. Thoroughly convinced of the practical superiority of the Hindu-Arabic methods of calculation, Fibonacci, in 1202, shortly after his return home, published his famous *Liber abaci*.

The first edition of the *Liber abaci* is no longer extant, and the

---

*The operations of canceling like terms occurring on the two sides of an equation, and of transposing a term from one side to the other by changing the sign of the term, constituted the basic rules of early algebra.

work is known to us through a second edition that came out in 1228. The work is devoted to arithmetic and algebra and, though essentially an independent investigation, shows the influence of the algebras of al-Khowârizmî and Abû Kâmil. The book profusely illustrates and strongly promotes the Hindu-Arabic notation and its accompanying computing algorithms. In the fifteen chapters of the work are explanations of the reading and writing of the new numerals, methods of calculation with integers and fractions, computation of square and cube roots, and the solution of linear and quadratic equations both by false position and by algebraic procedures. Negative and imaginary roots of equations are not recognized and the algebra is rhetorical. Applications are given involving barter, partnership, alligation, and mensurational geometry. The work also contains a large collection of problems which served later authors as a storehouse for centuries. But the immediate impact of the work was its effect on the promulgation of the Hindu-Arabic numeral system.

The nature of some of the problems found in the *Liber abaci* can be gathered from the exercises attached to this lecture. We shall here comment on two especially interesting problems of the work.

Little difficulty was encountered in deciphering and then interpreting most of the problems of the ancient Rhind papyrus (ca. 1650 B.C.), but there is one problem—Problem 79—for which the interpretation at first was baffling. The problem appears simply as the following set of data and indicated addition, here transcribed:

Estate

Houses	7
Cats	49
Mice	343
Heads of wheat	2401
Hekat measures	16807
	19607

One soon recognizes the numbers as the first five powers of 7, along with their sum. Because of this it was at first thought that perhaps the writer was introducing the symbolic terminology *houses, cats,* and so on, for *first power, second power,* and so on.

A much more plausible and interesting explanation, however, was furnished in 1907 by Moritz Cantor (1829-1920), the eminent German historian of mathematics. He saw in this problem an ancient forerunner of a problem that was popular in the Middle Ages and that was given by Fibonacci in his *Liber abaci* in the following form: "There are seven old women on the road to Rome. Each woman has seven mules; each mule carries seven sacks; each sack contains seven loaves; with each loaf are seven knives; and each knife is in seven sheaths. Women, mules, sacks, loaves, knives, and sheaths, how many are there in all on the road to Rome?" As a later and more familiar version of the same problem we have the Old English children's rhyme:

> As I was going to St. Ives
> I met a man with seven wives;
> Every wife had seven sacks;
> Every sack had seven cats;
> Every cat had seven kits.
> Kits, cats, sacks, and wives,
> How many were going to St. Ives?

According to Cantor's interpretation, the original problem in the Rhind papyrus might then be formulated as follows: "An estate consists of seven houses; each house had seven cats; each cat ate seven mice; each mouse ate seven heads of wheat; and each head of wheat was capable of yielding seven hekat measures of grain. Houses, cats, mice, heads of wheat, and hekat measures of grain, how many of these in all were in the estate?"

Here, then, may be a problem which has been preserved as part of the puzzle lore of the world. It was apparently already old when the scribe Ahmes copied it in the Rhind papyrus, and older by close to three thousand years when Fibonacci incorporated a version of it in his *Liber abaci*. Almost eight hundred years later we are reading another variant of it to our children. One cannot help wondering if a surprise twist such as occurs in the Old English rhyme may also have occurred in the ancient Egyptian problem, though, in all likelihood, this twist was an Anglo-Saxon contribution.

There are many puzzle problems popping up every now and then in our present-day magazines which have medieval counterparts.

How much further back some of them go is now almost impossible to determine.*

The second particularly interesting problem in the *Liber abaci* that we want to mention, and probably the most fruitful problem in the work, is the following: "How many pairs of rabbits can be produced from a single pair in a year if every month each pair begets a new pair which from the second month on becomes productive?" Without much effort, one can show that this problem leads to the following interesting sequence (wherein the terms are the number of pairs of rabbits present in successive months):

$$1, 1, 2, 3, 5, \ldots, x, y, x + y, \ldots .$$

This sequence, in which the first two terms are 1's and then each succeeding term is the sum of the two immediately preceding ones, has become known as the *Fibonacci sequence* and it occurs in an astonishing number of unexpected situations. It has applications to dissection puzzles (like the dissection of a square into unequal squares), to art, to the propagation of bees, and to phyllotaxis, and it appears surprisingly in various places in mathematics.

Consider, for example, the seed head of a sunflower. The seeds are found in small diamond-shaped pockets bounded by arcs of spiral curves radiating from the center of the head to the outside edge, as illustrated in Figure 53. Now a curious thing is this, if one should count the number of clockwise spirals and then the number of counterclockwise spirals, these two numbers will be found to be successive terms in the Fibonacci sequence. Indeed, this is true of the seed head of any composite flower (for instance, a daisy, or an aster); it is more easily tested on a sunflower, because these flowers have such large seeds and seed heads. Incidentally, as a further curiosity, the above-mentioned spirals are logarithmic spirals.

Next consider the leaves (or buds, or branches) growing out of the side of a stalk of a plant. If we fix our attention on some leaf near the bottom of the stalk and then count the number of leaves up the stalk until we come to one that is directly over the original leaf, this number is generally a term of the Fibonacci sequence. Also, as we

---

*See D. E. Smith, "On the Origin of Certain Typical Problems," *American Mathematical Monthly*, vol. 24 (February 1917), pp. 64–71.

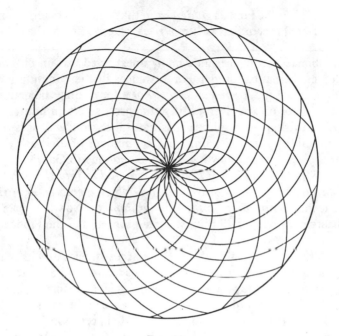

FIG. 53

ascend the stalk and count the number of times we revolve about the
stalk before we come to the leaf directly over the original one, this
number is generally the preceding alternate term of the sequence.
Similar arrangements occur in a wide variety of plant forms, such as
the leaves of a head of lettuce, in the layers of an onion, and the conical spirals of a pine cone.

If we form the sequence of ratios of successive terms of the
Fibonacci sequence, we obtain

$$\frac{1}{1}, \frac{1}{2}, \frac{2}{3}, \frac{3}{5}, \frac{5}{8}, \frac{8}{13}, \ldots .$$

It can be shown mathematically that this sequence of ratios approaches the number

$$r = (\sqrt{5} - 1)/2$$

as a limit. This is the famous *golden ratio* that was considered in LEC-
TURE 5. It would seem that nature strives to approximate the golden
ratio *r*.

The Fibonacci sequence finds many unexpected uses in different
parts of mathematical study. For instance, in the computational
procedure called the *Euclidean algorithm,** for finding the greatest
common divisor of two given positive integers, a certain number of
successive divisions is required. It is natural to wonder if it is possible to establish *a priori* a limit for the number of divisions. The
answer is given by the following neat theorem due to Gabriel Lamé
(1795–1870): *The number of divisions required to find the greatest
common divisor of two positive integers is never greater than five
times the number of digits in the smaller number.* Now the proof of
this theorem utilizes, of all things, some properties of the Fibonacci
sequence!

The literature on the ubiquitous Fibonacci sequence and its many
properties is incredibly large and continues to grow. The interesting
relations seem, like the geometry of the triangle, to be inexhaustible.
In fact, in 1963, a group of Fibonacci-sequence enthusiasts, headed
by Dr. Verner Hoggatt, Jr., founded the Fibonacci Association and
began publication of a journal, *The Fibonacci Quarterly,* devoted
principally to research on the Fibonacci and allied sequences. In its
first three years of existence, this journal published close to 1000
pages of research in this particular field. In 1968, three extra issues
of the journal appeared in a desperate effort to catch up somewhat
on the large manuscript backlog. This feverish activity has continued unabated.

Fibonacci wrote other works besides his *Liber abaci.* Thus, in
1220 appeared his *Practica geometriae,* a vast collection of material
on geometry and trigonometry treated skillfully, rigorously, and
with originality. About 1225 Fibonacci wrote his *Liber quadratorum,* a brilliant and novel work on indeterminate analysis, which
has marked him as the outstanding mathematician in the field between Diophantus and Fermat. These works were quite beyond the
abilities of most of the contemporary scholars.

Fibonacci's mathematical talent came to the attention of the

---

*See Exercise 8.4 of LECTURE 8.

patron of learning Emperor Frederick II of the Norman kingdom of
Sicily, with the result that Fibonacci was invited to court to par-
ticipate in a mathematical tournament. Three problems were set by
John of Palermo, a member of the king's retinue. Fibonacci solved
all three problems, a performance that evoked considerable admiration.

The first problem was to find a rational number x such that $x^2 +$
5 and $x^2 - 5$ shall each be a square of a rational number. Fibonacci
gave the answer $x = 41/12$, which is correct, since $(41/12)^2 + 5 =$
$(49/12)^2$ and $(41/12)^2 - 5 = (31/12)^2$. The solution appeared later
in the *Liber quadratorum.*

The second problem was to find a solution of the cubic equation

$$x^3 + 2x^2 + 10x = 20.$$

Fibonacci attempted a proof that no root of the equation can be expressed by means of irrationalities of the form $\sqrt{a} + \sqrt{b}$, or, in
other words, that no root can be constructed with straightedge and
compasses. He then obtained an approximate answer, which, expressed in decimal notation, is

$$1.3688081075,$$

and is correct to nine places. The answer later appeared, without
any accompanying explanation, in a work by Fibonacci entitled the
Flos ("blossom" or "flower") and has excited wonder as to how
Fibonacci found it.

The third problem, which is also recorded in the *Flos,* is the
easiest of the three, and was as follows: "Three men possess a pile of
money, their shares being 1/2, 1/3, 1/6. Each man takes some
money from the pile until nothing is left. The first man returns 1/2
of what he took, the second 1/3, and the third 1/6. When the total
so returned is divided equally among the men it is found that each
then possesses what he is entitled to. How much money was in the
original pile, and how much did each man take from the pile?" Here
is essentially Fibonacci's solution of the problem. Let s denote the
original sum and $3x$ the total sum returned. Before each man received a third of the sum returned, the three possessed $s/2 - x$, $s/3$
$- x$, $s/6 - x$. Since these are the sums they possessed after putting
back 1/2, 1/3, 1/6 of what they had first taken, the amounts first

taken were $2(s/2 - x)$, $(3/2)(s/3 - x)$, $(6/5)(s/6 - x)$, and these amounts added together equal s. This yields $7s = 47x$, and the problem is indeterminate. Fibonacci took $s = 47$ and $x = 7$. Then the sums taken by the men from the original pile are 33, 13, 1.

Fibonacci sometimes signed his work with *Leonardo Bigollo*. Now *bigollo* has more than one meaning; it means both "traveler" and "blockhead." In signing his work as he did, Fibonacci may have meant that he was a great traveler, for so he was. But a story has circulated that he took pleasure in using this signature because his contemporaries considered him a blockhead (for his interest in the new Hindu-Arabic numerals), and it pleased him to show these critics what a blockhead could accomplish.

Exercises

15.1. (a) Show that when the sum of the digits of a natural number is divided by 9, one obtains the same remainder as when the number itself is divided by 9.

The act of obtaining the remainder when a given natural number is divided by an integer n is known as *casting out n's*. The theorem above shows that it is particularly easy to cast out 9's.

(b) Let us call the remainder obtained when a given natural number is divided by 9 the *excess* for that number. Prove the following two theorems: (1) *The excess for a sum is equal to the excess for the sum of the excesses of the addends.* (2) *The excess for the product of two numbers is equal to the excess for the product of the excesses of the two numbers.*

These two theorems furnish the basis for checking addition and multiplication by casting out 9's.

(c) Add and then multiply 478 and 993, and check by casting out 9's.

15.2. (a) Show that if the order of the digits of a natural number are permuted in any way to form a new number, then the difference between the old and the new numbers is divisible by 9.

This furnishes the basis for the *bookkeeper's check*. If the sums of the debit and credit entries in double-entry bookkeeping do not balance, and the difference between the two sums is divisible by 9, then it is quite likely that the error is due to a transposition in digits made when transcribing a debit or a credit into the book.

(b) Explain the following number trick: Someone is asked to think of a number; form a new number by reversing the order of the digits; subtract the smaller from the larger number; multiply the difference by any number whatever; scratch out any nonzero digit in the product and announce what is left. The conjurer finds the scratched-out digit by calculating the excess for the announced result and then subtracting this excess from 9.

15.3. One of the oldest methods for approximating the real roots of an equation is the rule known as *regula duorum falsorum*, often called the rule of *double false position*. This method seems to have originated in India and was used by the Arabians. In brief, and in modern form, the method is this: Let x_1 and x_2 be two numbers lying close to and on each side of a root x of the equation $f(x) = 0$. Then the intersection with the x-axis of the chord joining the points $(x_1, f(x_1))$, $(x_2, f(x_2))$ gives an approximation x_3 to the sought root (see Figure 54). Show that

$$x_3 = \frac{x_2 f(x_1) - x_1 f(x_2)}{f(x_1) - f(x_2)}.$$

The process can now be applied with the appropriate pair x_1, x_3 or x_3, x_2.

15.4. (a) Compute, by double false position, to three decimal places, the root of $x^3 - 36x + 72 = 0$ which lies between 2 and 3.

(b) Compute, by double false position, to three decimal places, the root of $x - \tan x = 0$ which lies between 4.4 and 4.5.

15.5. Obtain the Fibonacci sequence from the problem in the *Liber abaci* concerning the propagation of rabbits.

15.6. If u_n represents the nth term of the Fibonacci sequence show that

(a) $u_{n+1}u_{n-1} = u_n^2 + (-1)^n$, $n \geq 2$.
(b) $u_n = [(1 + \sqrt{5})^n - (1 - \sqrt{5})^n]/2^n\sqrt{5}$.
(c) $\lim_{n \to \infty}(u_n/u_{n+1}) = (\sqrt{5} - 1)/2$.
(d) u_n and u_{n+1} are relatively prime.

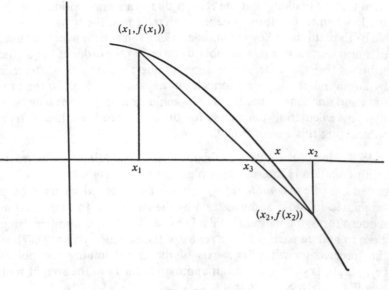

FIG. 54

15.7. Solve the following problems found in the *Liber abaci*. The first one was posed to Fibonacci by a magister in Constantinople; the second was designed to illustrate the rule of three.

(a) If A gets from B 7 denarii, then A's sum is fivefold B's; if B gets from A 5 denarii, then B's sum is sevenfold A's. How much has each?

(b) A certain king sent 30 men into his orchard to plant trees. If they could set out 1000 trees in 9 days, in how many days would 36 men set out 4400 trees?

15.8. Solve the following problem from the *Liber abaci;* it is an example of an inheritance problem which appeared in later works by Chuquet and Euler.

A man left to his oldest son one bezant and a seventh of what was left; then, from the remainder, to his next son he left two bezants and a seventh of what was left; then, from the new remainder, to his third son he left three bezants and a seventh of what was left. He continued this way, giving each son one more bezant than the previous son and a seventh of what remained. By this division it developed that the last son received all that was left and all the sons shared equally. How many sons were there and how large was the man's estate?

15.9. Show that the squares of the numbers

$$a^2 - 2ab - b^2, \quad a^2 + b^2, \quad a^2 + 2ab - b^2$$

are in arithmetic progression. If $a = 5$ and $b = 4$, the common difference is 720, and the first and third squares are $41^2 - 720 = 31^2$ and $41^2 + 720 = 49^2$. Dividing by 12^2 we obtain Fibonacci's solution to the first of the tournament problems. The problem is insolvable if the 5 in the problem is replaced by 1, 2, 3, or 4. Fibonacci showed that if x and h are integers such that $x^2 + h$ and $x^2 - h$ are perfect squares, then h must be divisible by 24. As examples we have $5^2 + 24 = 7^2$, $5^2 - 24 = 1$ and $10^2 + 96 = 14^2$, $10^2 - 96 = 2^2$.

15.10. Solve the following problem given by Fibonacci in his *Liber abaci*. This problem reappeared in a remarkable number of variations. It contains the essence of the idea of an annuity.

A man entered an orchard through seven gates, and there took a certain number of apples. When he left the orchard he gave the first guard half the apples that he had and one apple more. To the second guard he gave half his remaining apples and one apple more. He did the same to each of the remaining five guards, and left the orchard with one apple. How many apples did he gather in the orchard?

Further Reading

HOGGATT, V. E., Jr., *Fibonacci and Lucas Numbers.* Boston: Houghton Mifflin, 1969.

SULLIVAN, J. W. N., *The History of Mathematics in Europe from the Fall of Greek Science to the Rise of the Conception of Mathematical Rigour.* New York: Oxford University Press, 1925.

AN EXTRAORDINARY AND BIZARRE STORY

In LECTURE 14 we saw how the Persian poet-mathematician Omar Khayyam solved cubic equations geometrically. In this lecture we shall see how, almost 500 years later, Italian mathematicians finally managed to solve cubic equations, and then, shortly after, quartic equations also, algebraically. These accomplishments constitute spectacular episodes in the history of mathematics and yield two GREAT MOMENTS IN MATHEMATICS, one following closely upon the heels of the other. There are elements of the colorful and of the bizarre in the story, and some of the characters in the tale are among the most unusual in all of mathematics.

Briefly told, the facts seem to be these. About 1515, Scipione del Ferro (1465–1526), a professor of mathematics at the University of Bologna, succeeded in algebraically solving cubic equations lacking the quadratic term, that is, cubic equations of the form $x^3 + mx = n$, probably basing his work on earlier Arabic sources. He did not publish his discovery, but revealed it as a secret to his pupil Antonio Maria Fior. Now, about 1535, Nicolo Fontana (ca. 1499–1557), commonly referred to as Tartaglia (the stammerer) because of a childhood injury that affected his speech, claimed to have discovered an algebraic solution of cubic equations lacking the linear term, that is, cubic equations of the form $x^3 + px^2 = q$. Believing this claim to be a mere boastful bluff, Fior challenged Tartaglia to a public contest of solving, within a given time, a set of cubic equations to be submitted in equal numbers by the two contestants at the scheduled time of the contest. Accepting the challenge, Tartaglia exerted himself, and only a few days before the scheduled date found an algebraic solution of cubic equations lacking the quadratic term. Entering the contest able to solve two types of cubic equations, whereas Fior could solve but one type, Tartaglia triumphed completely. Later, Girolamo Cardano (1501–1576), an unprincipled genius who taught mathematics and practiced medicine in Milan, upon giving a solemn pledge of secrecy, wheedled the key to the solution of cubic equations from Tartaglia. In 1545, Cardano published, at Nuremberg, Germany, his *Ars magna,* a great Latin treatise on algebra, and as a crowning jewel he inserted Tartaglia's solution of cubic equations. Tartaglia vehemently protested, but his complaints were met by Ludovico Ferrari (1522–1565), Cardano's most capable pupil, who argued that Cardano had received his infor..1ation from del Ferro through a third party, and accused Tartaglia of plagiarism from the same source. There ensued an acrimonious dispute from which Tartaglia was perhaps lucky to escape with his life.

One finds variations in some of the details of this story, due to the fact that the actors in the drama seem not always to have had the highest regard for truth.

The algebraic solution of the cubic equation $x^3 + mx = n$ given by Cardano in his *Ars magna* is essentially the following. Consider the identity

$$(a - b)^3 + 3ab(a - b) = a^3 - b^3.$$

If we choose a and b so that

$$3ab = m, \qquad a^3 - b^3 = n,$$

then x is given by $a - b$. But, solving the last two equations for a and b, one finds

$$a = \sqrt[3]{(n/2) + \sqrt{(n/2)^2 + (m/3)^3}},$$
$$b = \sqrt[3]{-(n/2) + \sqrt{(n/2)^2 + (m/3)^3}},$$

and x is then determined by the so-called *Cardano-Tartaglia formula*

$$x = \sqrt[3]{(n/2) + \sqrt{(n/2)^2 + (m/3)^3}}$$

$$- \sqrt[3]{-(n/2) + \sqrt{(n/2)^2 + (m/3)^3}}.$$

Since, by the transformation $x = z - b/3a$, the general cubic equation

$$ax^3 + bx^2 + cx + d = 0$$

can be converted into one of the form

$$z^3 + mz = n,$$

the solution above suffices for solving all cubic equations.

It was not long after the general cubic had been solved algebraically that an algebraic solution was discovered for the general quartic (or biquadratic) equation. In 1540, the Italian mathematician Zuanne de Tonini da Coi proposed the following problem to Cardano: "Divide 10 into three parts such that they shall be in continued proportion and that the product of the first two shall be 6." If the three parts be denoted by a, b, c, we have

$$a + b + c = 10, \qquad ac = b^2, \qquad ab = 6.$$

It is straightforward to show that elimination of a and c yields the quartic equation

$$b^4 + 6b^2 + 36 = 60b.$$

Cardano was unable to solve the resulting equation, but his pupil Ferrari succeeded, and in so doing devised a method for solving any quartic equation of the form

$$x^4 + px^2 + qx + r = 0,$$

a form to which any quartic is easily reduced by a simple linear transformation. From the above quartic one obtains

$$x^4 + 2px^2 + p^2 = px^2 - qx - r + p^2$$

or

$$(x^2 + p)^2 = px^2 - qx + p^2 - r,$$

whence, for arbitrary y,

$$(x^2 + p + y)^2 = px^2 - qx + p^2 - r + 2y(x^2 + p) + y^2$$
$$= (p + 2y)x^2 - qx + (p^2 - r + 2py + y^2).$$

Now let us choose y so that the right member of the equation above is a square. This is the case when*

$$4(p + 2y)(p^2 - r + 2py + y^2) - q^2 = 0.$$

But this is a cubic in y and may therefore be solved by previous methods. Such a value of y reduces the original problem to nothing but the extraction of square roots.

Cardano had the pleasure of also incorporating Ferrari's solution of quartic equations in his *Ars magna* of 1545.

Other algebraic solutions of the general cubic and quartic equations soon appeared. Toward the end of our lecture we shall note the methods (posthumously published in 1615) devised by the sixteenth-century mathematician François Viète (1540-1603), and also a method for solving quartic equations that was given in 1637 by René Descartes (1596-1650). Right now, however, for interest's sake, we give some biographical details about the two principals—Cardano and Tartaglia—in the story of the algebraic solution of cubic equations.

Girolamo Cardano, one of the most extraordinary characters in the history of mathematics, was born in Pavia in 1501 as the illegitimate son of a jurist. He developed into a man of passionate contrasts, starting his professional life as a physician while studying, teaching, and writing mathematics on the side. After a trip to Scotland, he successively held prominent chairs at the universities of Pavia and Bologna. As a practicing astrologer he cast horoscopes, and at one time was imprisoned for heresy because of his audacity in publishing a horoscope of Christ's life. Resigning his chair at Bologna, he moved to Rome and became a distinguished astrologer, curiously, in view of his earlier imprisonment for heresy, receiving a pension as astrologer to the papal court. He died in Rome in 1576, by his own hand, one report says, so as to fulfill his own earlier astrological prediction of the time of his death. Many stories are told of his violent temper, as when, in a fit of rage, he cut off the ears of his younger son because the boy had

---

*A necessary and sufficient condition for the quadratic $Ax^2 + Bx + C$ to be the square of a linear function is that $4AC - B^2 = 0$.

been disturbingly noisy. He had many enemies, and it could be that some of the stories are exaggerations circulated by those whom he had crossed. Thus it may be that he has been overly maligned; this viewpoint, of course, is supported by his autobiography.

As one of the most talented and many-sided men of his day, Cardano wrote several works on arithmetic, astronomy, physics, medicine, and other subjects. His greatest work is his *Ars magna*, which was the first eminent Latin work devoted solely to algebra. In it notice is taken of negative roots of an equation, and some attention is given to computations with imaginary numbers. There is also a crude procedure for finding approximate values of real roots of polynomial equations, and there is some evidence that he was familiar with "Descartes' rule of signs," sometimes encountered by a student of today in his college algebra course. As an inveterate gambler, Cardano wrote a gambler's manual in which occur some interesting questions on probability.

Tartaglia endured a hard and impoverished childhood. He was born in Brescia about 1499 and was present as a young boy at the taking of Brescia by the French in 1512. During the brutalities that accompanied the taking of Brescia, Tartaglia and his father (who was a postal messenger of the town) fled with many others into the cathedral for sanctuary. The French soldiers, however, pursued, and a massacre took place within the holy walls. The father was slain, and the boy, with a split skull and a severe saber cut that cleft his jaws and palate, was left for dead. When the boy's mother later reached the cathedral to search for her family, she found her husband dead but the son still alive. She managed to carry off the boy, and, lacking resources for a doctor's assistance, recalled that a wounded dog always licks the injured spot. So the mother licked the boy's wounds. Tartaglia later attributed his recovery to this primitive therapy. But the injury to his palate caused a lifelong imperfection in his speech, from which he received his nickname of "the stammerer." His mother managed to gather together enough money to send him to school for 15 days, and he made the most of this small opportunity by stealing a copybook from which he subsequently taught himself how to read and write. Lacking the means to buy writing materials, it is said that he hied himself to the cemetery to use the tombstones as slates. Later in life he earned his livelihood by teaching science and mathematics in various cities of Italy. He died in Venice in 1557.

Tartaglia was a gifted mathematician. In addition to his work on cubic equations, he was perhaps the first to apply mathematics to the science of artillery fire. He wrote what is generally considered the best Italian arithmetic of the sixteenth century. This was a two-volume work containing a full treatment of the numerical operations and the commercial customs of his time. He also published editions of Euclid and Archimedes.

In 1572, a few years before Cardano died, Rafael Bombelli (ca. 1526-1573) published an algebra which contributed further to our knowledge of cubic equations. In textbooks on the theory of equations, it is shown that, when the three roots of the cubic equation $x^3 + mx - n$ are all real and different from zero, the expression

$$(n/2)^2 + (m/3)^3$$

is negative. But the Cardano-Tartaglia formula then expresses the real roots of the cubic equation as the difference of two cube roots of *complex imaginary numbers*. Imaginary numbers were little understood at the time, and the anomalous situation of expressing real roots in terms of imaginary numbers considerably bothered the algebraists of the time. Bombelli pointed out that, using modern terminology, just as the two radicands are *conjugate* imaginary numbers, so also must the two cube roots be conjugate imaginary numbers, and thus their sum turns out to be a real number. But any attempt to find algebraically the two cube roots of the imaginary numbers presented by the Cardano-Tartaglia formula merely leads to the very cubic equation whose solution is being sought in the first place. Because of this impasse, which arises whenever all three roots are nonzero real numbers, this situation became known as the "irreducible case." Here one has an expression for the real roots of the cubic equation, but the form of the expression is useless for practical purposes. The impasse was later circumvented by means of a trigonometrical approach.

It will be recalled from LECTURE 12 that Bombelli assisted in the development of algebraic notation.

Other, and somewhat different, algebraic solutions of cubic and quartic equations were furnished by later writers. Thus François Viète, in the 1615 posthumous publication, gave the following elegant solution of the cubic equation

$$x^3 + 3ax = 2b,$$

a form to which any cubic can be reduced. Setting

$$x = \frac{a}{y} - y$$

the cubic equation becomes

$$y^6 + 2by^3 = a^3,$$

a quadratic in y^3. We thus find y^3, and then y, and then x. Viète's solution of the quartic equation is similar to Ferrari's. Consider the "depressed" quartic

$$x^4 + ax^2 + bx = c,$$

a form to which any quartic can be reduced. This may be written as

$$x^4 = c - ax^2 - bx.$$

Adding $x^2y^2 + y^4/4$ to both sides yields

$$\left(x^2 + \frac{y^2}{2}\right)^2 = (y^2 - a)x^2 - bx + \left(\frac{y^4}{4} + c\right).$$

Now choose y so that the right member is a perfect square. The condition for this is

$$y^6 - ay^4 + 4cy^2 = 4ac + b^2,$$

a cubic in y^2. Such a y^2 may be found and the problem completed by extracting square roots.

Descartes' 1637 solution of a depressed quartic equation

$$x^4 + bx^2 + cx + d = 0$$

employs the *method of undetermined coefficients*. Set the left member of the equation equal to the product

$$(x^2 + kx + h)(x^2 - kx + m).$$

By equating corresponding coefficients on the two sides of the resulting equation, one obtains three relations connecting k, h, m. Eliminating h and m from the three relations yields a sextic equation in k which can be regarded as a cubic equation in k^2. Thus the solution of the original quartic equation is reduced to the solution of an associated cubic equation.

Since the solution of the general quartic equation can be made to depend on the solution of an associated cubic equation, Euler, about 1750, attempted similarly to reduce the solution of the general quintic equation to that of an associated quartic equation. He failed in this attempt, as did Lagrange about thirty years later. An Italian physician, Paola Ruffini (1765-1822), in 1803, 1805, and 1813 supplied an inconclusive proof of what is now known to be a fact, that the roots of a general fifth, or higher, degree equation cannot be expressed by means of radicals in term of the coefficients of the equation. This remarkable fact was independently established later, in 1824, by the famous Norwegian mathematician Niels Henrik Abel (1802-1829). Évariste Galois (1811-1832), who died in a pistol duel when he was only 21 years old, left behind him, in the form of a scientific testament, a letter which, when finally unraveled, was found, among other things, to supply criteria for the possibility of solving an algebraic equation by radicals. But all this belongs to a more recent GREAT MOMENT IN MATHEMATICS.

Exercises

16.1.(a) Show that the transformation $x = z - a_1/na_0$ converts the n-ic equation

$$a_0x^n + a_1x^{n-1} + a_2x^{n-2} + \cdots + a_n = 0$$

into an equation in z which lacks the $(n - 1)$st degree term.

(b) By part (a) the transformation $x = z - b/3a$ converts the general cubic equation

$$ax^3 + bx^2 + cx + d = 0$$

into one of the form

$$z^3 + 3Hz + G = 0.$$

Find H and G in terms of a, b, c, d.

16.2. Carry out the steps in the Cardano-Tartaglia solution of the cubic equation

$$x^3 + mx = n.$$

16.3. Solve $x^3 + 63x = 316$, for one root, by the Cardano-Tartaglia formula.

16.4. Show that da Coi's problem of 1540 leads to the quartic equation

$$x^4 + 6x^2 + 36 = 60x.$$

16.5. Obtain, by Ferrari's method, the cubic equation associated with the quartic equation of Exercise 16.4.

16.6. As an example of the irreducible case in cubics solve

$$x^3 - 63x = 162$$

by the Cardano-Tartaglia formula. Then show that

$$(-3 + 2\sqrt{-3})^3 = 81 + 30\sqrt{-3} \quad \text{and} \quad (-3 - 2\sqrt{-3})^3$$
$$= 81 - 30\sqrt{-3},$$

whence the root given by the formula is -6 in disguise.

16.7. Cardano solved the particular quartic equation

$$13x^2 = x^4 + 2x^3 + 2x + 1$$

by adding $3x^2$ to both sides. Do this and solve the equation for all four roots.

16.8. Solve $x^3 + 63x = 316$ by Viète's method.

16.9. Obtain, by Viète's method, the cubic equation associated with the quartic equation

$$x^4 + 6x^2 + 36 = 60x.$$

16.10.(a) Carry out the details of Descartes' 1637 solution of the depressed quartic equation

$$x^4 + bx^2 + cx + d = 0.$$

(b) Find, by Descartes' method, the cubic equation associated with the quartic equation

$$x^4 - 2x^2 + 8x - 3 = 0.$$

Given that one root of the associated cubic is 4, obtain the four roots of the quartic equation.

Further Reading

CARDAN, JEROME, *The Book of My Life*, tr. by Jean Stoner. New York: Dover, 1963.

ORE, OYSTEIN, *Cardano, the Gambling Scholar*. Princeton, N.J.: Princeton University Press, 1953.

SULLIVAN, J. W. N., *The History of Mathematics in Europe from the Fall of Greek Science to the Rise of the Conception of Mathematical Rigour*. New York: Oxford University Press, 1925.

DOUBLING THE LIFE OF THE ASTRONOMER

In 1614, a Scottish nobleman, living in Edinburgh, published the details of a wonderful invention that he had made. News of the invention traveled fast, and, in the following year, after some correspondence, a professor of mathematics left London by horse-drawn carriage for the long trip to Edinburgh so that he might pay his personal respects to the ingenious Scotsman. On the interminable journey over the bumpy dirt roads, the professor recorded some of his thoughts in his diary. How tall a forehead, he ruminated, must the nobleman possess in order to house the brains sufficient for discovering so remarkable an invention. Unanticipated delays prolonged the time of the journey, and the disappointed awaiting nobleman at Edinburgh finally complained to a friend. "Ah, John," he said, "the professor will not come." At that very moment a knock was heard at the gate and the professor was ushered into the nobleman's presence. For almost a quarter of an hour each man beheld the other without speaking a word. Then the professor said, "My lord, I have undertaken this long journey purposely to see your person, and to learn by what engine of wit or ingenuity you came first to think of this most excellent help in astronomy. But, my lord, being by you found out, I wonder nobody found it out before when now known it appears so easy." The professor remained for a month as an honored guest at the nobleman's castle.

The Scottish nobleman was John Napier (1550–1617) of Merchiston Castle at Edinburgh; the mathematician was Henry Briggs (1561–1631), professor of geometry at Gresham College in London; the remarkable invention was that of logarithms, one of the great labor-saving devices in the field of computation and decidedly marking a GREAT MOMENT IN MATHEMATICS.

As the high school mathematics student learns today, the power of logarithms as a computing device lies in the fact that by them multiplication and division are reduced to the simpler operations of addition and subtraction. A forerunner of this reduction is seen in the formula

$$\sin A \sin B = \tfrac{1}{2}[\cos (A - B) - \cos (A + B)],$$

which was well known in Napier's time, and it is quite likely that Napier's line of thought started with this formula, since otherwise it is difficult to account for his initial restriction of logarithms to those of the sines of angles. Napier labored more than twenty years upon his theory, and, whatever its genesis, his final definition of a logarithm is as follows. Consider a line segment AB and an infinite ray DE, as shown in Figure 55. Let points C and F start moving simultaneously from A and D, respectively, along these lines, with the same initial rate. Suppose C moves with a velocity always numerically equal to the distance CB, and that F moves with a uniform velocity. Then Napier defined DF to be the logarithm of CB. That is, setting $DF = x$ and $CB = y$,

$$x = \text{Nap log } y.$$

To avoid the nuisance of fractions, Napier took the length of AB as 10^7, for the best table of sines available to him extended to seven places. From Napier's definition, and through knowledge not available in Napier's day, it develops that*

---

*The result is easily shown with the aid of a little calculus. Thus we have $AC = 10^7 - y$, whence

$$\text{velocity of } C = -dy/dt = y.$$

That is, $dy/y = -dt$, or, integrating, $\ln y = -t + C$. Evaluating the constant of integration by substituting $t = 0$, we find that $C = \ln 10^7$, whence

$$\ln y = -t + \ln 10^7.$$

Now

$$\text{velocity of } F = dx/dt = 10^7,$$

so that $x = 10^7 t$. Therefore

$$\text{Nap log } y = x = 10^7 t = 10^7 (\ln 10^7 - \ln y)$$
$$= 10^7 \ln(10^7/y) = 10^7 \log_{1/e} (y/10^7).$$

$$\text{Nap } \log y = 10^7 \log_{1/e} (y/10^7),$$

so that the frequently made statement that Napierian logarithms are natural logarithms (that is, logarithms to the base e) is actually not true. One observes that the Napierian logarithm *decreases* as the number increases, just opposite to what happens with natural logarithms.

Napier published his discussion of logarithms in 1614 in a brochure entitled *Mirifici logarithmorum canonis descriptio* (A Description of the Wonderful Law of Logarithms). The work contains a table giving the Napierian logarithms of the sines of angles for successive minutes of arc. The *Descriptio* roused immediate and widespread interest. It was when Briggs visited Napier in 1615 that both men agreed that the tables would be more useful if they were altered so that the logarithm of 1 would be 0 and the logarithm of 10 would be an appropriate power of 10. Thus were born the so-called *Briggsian*, or *common*, logarithms taught in the high schools of today. These logarithms, which are essentially logarithms to the base 10, owe their superior utility in numerical computations to the fact that our number system also is based on 10. For a number system having some other base b it would, of course, be more convenient for computational purposes in that system to have tables of logarithms also to the base b.*

Upon his return to London after his visit with Napier, Briggs devoted all his energies toward the construction of a table of common logarithms and in 1624 published his *Arithmetica logarithmica*, containing a 14-place table of common logarithms of the numbers from 1 to 20,000 and from 90,000 to 100,000. The gap

---

*For theoretical and analytical purposes, the best base for a system of logarithms is that which yields the simplest possible form for the derivative of the logarithm function. In calculus texts it is shown that if $y = \log_b x$, then

$$dy/dx = (1/x) \log_b e.$$

If we choose $b = e$, this derivative assumes its simplest form, namely,

$$dy/dx = 1/x.$$

Logarithms to the base e are called *natural logarithms* and the special symbolism $\ln x$ has been generally adopted for $\log_e x$.

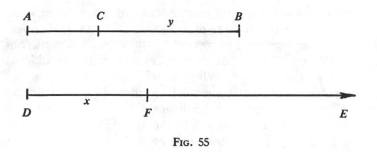

FIG. 55

from 20,000 to 90,000 was later filled in, with assistance, by Adriaen Vlacq (1600-1666), a Dutch bookseller and publisher. Earlier, in 1620, Edmund Gunter (1581-1626), professor of astronomy at Gresham College and a colleague of Briggs, had published a seven-place table of the common logarithms of the sines and tangents of angles for intervals of a minute of arc. It was Gunter who coined the words *cosine* and *cotangent;* he is known to surveyors for his "Gunter's chain." Briggs and Vlacq published four fundamental tables of logarithms, which were not superseded until, between 1924 and 1949, extensive 20-place tables were calculated in England as part of the celebration of the tercentenary of the invention of logarithms.

The word *logarithm* means "ratio number," and was adopted by Napier after first employing the expression *artificial number.* Briggs introduced the word *mantissa* for the decimal part of the logarithm. This word is a late Latin term of Etruscan origin, originally meaning an "addition" or "makeweight," and which in the sixteenth century came to mean "appendix." The term *characteristic,* for the integral part of the logarithm, was also suggested by Briggs and was used by Vlacq. It is curious that it was customary in early tables of common logarithms to print the characteristic as well as the mantissa, and it was not until the eighteenth century that the present practice of printing only the mantissas was established.

Napier's remarkable invention was enthusiastically adopted throughout Europe. In astronomy, in particular, the time was over-ripe for such a discovery; as Laplace asserted, the invention of logarithms "by shortening the labors doubled the life of the astronomer." Johann Kepler, about whom we shall have more to say in our next lecture, did much to bring logarithms into vogue in Germany. A similar service was rendered by Bonaventura Cavalieri in Italy and Edmund Wingate in France. Cavalieri will be considered more fully in LECTURE 19; Wingate, who spent many years in France, became the most prominent seventeenth-century British writer of elementary arithmetic textbooks.

At times in mathematics, as will be exemplified in a later lecture when we come to discuss the discovery of non-Euclidean geometry, a particular mathematical invention or discovery occurs more or less simultaneously to more than one person, as though the time had become ripe for the invention and it simply had to burst forth. In the case of logarithms, however, Napier's only serious rival appears to have been the Swiss watchmaker Jobst Bürgi (1552–1632). Bürgi conceived and constructed a table of logarithms independently of Napier, publishing his results in 1620, six years after Napier had published his. Each man, of course, had conceived the idea of logarithms long before publishing, but it is generally believed that Napier had the idea first. The approaches of the two men, however, were quite different, for Napier's was geometric and Bürgi's was algebraic. Today a logarithm is universally regarded as an exponent. Thus if $n = b^x$, we say x is the logarithm of n to the base b, and we write $x = \log_b n$. With this definition, the laws of logarithms are nothing but a restatement of the laws of exponents. One of the oddities in the history of mathematics is the fact that logarithms were discovered before exponents were in use.

Napier lived most of his life at the imposing family estate of Merchiston Castle, near Edinburgh, Scotland, having been born there in 1550 when his father was only 16 years old, and having died there in 1617, only three years after announcing his great discovery to the world.

Napier devoted much of his time and energy to the political and religious controversies of his day. He was vehemently anti-Catholic and espoused the causes of John Knox and James I. In 1593 he published an acid and widely read attack on the Church of Rome entitled *A Plaine Discouery of the whole Reuelation of Saint Iohn,* in which he attempted to show that the Pope was anti-Christ and that the Creator intended to end the world in the years between 1688 and

1700. The book was immensely popular and ran through twenty-one editions, at least ten of them during Napier's lifetime; and Napier genuinely believed that his reputation with posterity would rest upon this book. How wrong he proved to be. His book is today totally disregarded and known only to a curious few. Instead, his reputation today rests solidly, widely, and almost solely upon one of his mathematical diversions, the invention of logarithms.

Napier was also the science-fiction author of his day, writing prophetically of various infernal war engines, accompanying his writing with plans and diagrams. He predicted that the future would develop a piece of artillery that could "clear a field of four miles circumference of all living creatures exceeding a foot of height," that it would produce "devices for sayling under water," and that it would create a chariot with "a living mouth of mettle" that would "scatter destruction on all sides." In World War I these were realized as the machine gun, the submarine, and the army tank.

As relaxation from his political and religious polemics, Napier amused himself with the study of mathematics and science, with the result that four products of his genius are now recorded in the history of mathematics. These are: (1) the invention of logarithms; (2) a clever mnemonic, known as the *rule of circular parts,* for reproducing the ten formulas used in solving right spherical triangles; (3) at least two trigonometric formulas of a group of four known as *Napier's analogies,* useful in the solution of oblique spherical triangles; and (4) the invention of a device, called *Napier's rods,* or *Napier's bones,* used for mechanically multiplying, dividing, and extracting square roots of numbers. The interested student can find a brief discussion of the last three in the exercises accompanying this lecture.

Henry Briggs, who did so much to advance logarithms, had the honor of occupying, for a time, the first professorial chair of mathematics established in Great Britain, a chair in geometry founded by Sir Thomas Gresham in 1596 at Gresham College in London; indeed, Briggs was the first incumbent of this chair. Sir Henry Savile, one-time warden of Merton College at Oxford University, later provost of Eton, and a lecturer on Euclid at Oxford, in 1619 founded two professorial chairs at Oxford, one in geometry and one in astronomy. Briggs had the honor also to be the first occupant of the

Savilian chair of geometry at Oxford. Briggs published ten works during his lifetime and left six others unpublished. His published works include treatises on navigation, Euclid's *Elements*, logarithms, and trigonometry.

Edmund Gunter, who constructed the first table of common logarithms of sines and tangents of angles, devised the logarithmic scale, a line of numbers in which the distances from the left end of the scale are proportional to the logarithms of the numbers indicated, and mechanically performed multiplications and divisions by adding and subtracting segments of the scale with the aid of a pair of dividers. The idea of carrying out these additions and subtractions by having two like logarithmic scales, one sliding along the other as shown in Figure 56, is due to William Oughtred (1574-1660). Although Oughtred invented such a simple slide rule as early as 1622, he did not describe it in print until 1632.

For years logarithms have been taught in the late high school or the early college mathematics courses, and also for years the slide rule, hanging from the belt in a handsome leather case, was the badge of recognition of the engineering students of a university campus. Today, however, with the advent of the amazing and increasingly inexpensive little pocket calculators, no one in his right mind would use a table of logarithms or a slide rule for calculation purposes. The teaching of logarithms as a computing device is vanishing from the schools, the famous makers of precision slide rules are discontinuing their production, and noted handbooks of mathematical tables are beginning to abandon the inclusion of logarithm tables. The products of Napier's GREAT MOMENT IN MATHEMATICS have become museum pieces.

The logarithmic function, on the other hand, will never die, for the simple reason that logarithmic and exponential variations are a vital part of nature and of analysis. Accordingly, a study of the

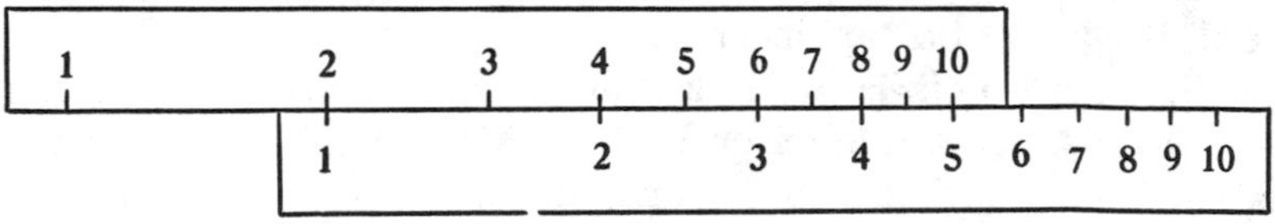

FIG. 56

mathematical properties of the logarithmic function, and of its inverse, the exponential function, will always remain an important part of mathematical instruction.

Exercises

17.1. Using the familiar laws of exponents, establish the following useful properties of logarithms:
    (a) $\log_b mn = \log_b m + \log_b n$
    (b) $\log_b(m/n) = \log_b m - \log_b n$
    (c) $\log_b(m^r) = r \log_b m$
    (d) $\log_b{}^s\sqrt{m} = (\log_b m)/s$

17.2. Show that
    (a) $\log_a N = \log_b N/\log_b a$ (With this formula one may compute logarithms to a base a when one has available a table of logarithms to base b.)
    (b) $\log_N b = 1/\log_b N$
    (c) $\log_N b = \log_{1/N}(1/b)$

17.3. By extracting the square root of 10, then the square root of the result thus obtained, and so on, the following table can be constructed:

$$
\begin{array}{ll}
10^{1/2} = 3.16228 & 10^{1/256} = 1.00904 \\
10^{1/4} = 1.77828 & 10^{1/512} = 1.00451 \\
10^{1/8} = 1.33352 & 10^{1/1024} = 1.00225 \\
10^{1/16} = 1.15478 & 10^{1/2048} = 1.00112 \\
10^{1/32} = 1.07461 & 10^{1/4096} = 1.00056 \\
10^{1/64} = 1.03663 & 10^{1/8192} = 1.00028 \\
10^{1/128} = 1.01815 & \cdots
\end{array}
$$

With this table one can compute the common logarithm of any number between 1 and 10, and hence, by adjusting the characteristic, of any positive number whatever. Thus, let N be any number between 1 and 10. Divide N by the largest number in the table which does not exceed N. Suppose the divisor is $10^{1/p_1}$ and the quotient is N_1. Then $N = 10^{1/p_1} N_1$. Treat N_1 in the same fashion, and continue the process, obtaining

$$N = 10^{1/p_1} 10^{1/p_2} \cdots 10^{1/p_n} N_n.$$

Stop when N_n differs from unity only in the sixth decimal place. Then, to five places,

$$N = 10^{1/p_1} 10^{1/p_2} \cdots 10^{1/p_n}$$

and

$$\log N = 1/p_1 + 1/p_2 + \cdots + 1/p_n.$$

This procedure is known as the *radix method* of computing logarithms.

(a) Compute log 4.26.

(b) Compute log 5.00.

17.4. There are ten formulas which are useful for solving right spherical triangles. There is no need to memorize these formulas, for it is easy to reproduce them by means of two rules devised by Napier. In Figure 57 is pictured a right spherical triangle, lettered in conventional manner. To the right of the triangle appears a circle divided into five parts, containing the same letters as the triangle, except C, arranged in the same order. The bars on c, B, A mean "the complement of" (thus $\bar{B}$ means $90° - B$). The angular quantities a, b, $\bar{c}$, $\bar{A}$, $\bar{B}$ are called the *circular parts*. In the circle there are two circular parts contiguous to any given part and two parts not contiguous to it. Let us call the given part the *middle part*, the two contiguous parts the *adjacent parts*, and the two noncontiguous parts the *opposite parts*. Napier's rules may be stated as follows:

1. The sine of any middle part is equal to the product of the cosines of the two opposite parts.

2. The sine of any middle part is equal to the product of the tangents of the two adjacent parts.

(a) By applying each of the above rules to each of the circular parts, obtain the ten formulas used for solving right spherical triangles.

(b) The formula connecting the sides a, b, c of a right spherical triangle is called the *Pythagorean relation* for the triangle. Find the Pythagorean relation for a right spherical triangle.

17.5. The following formulas are known as *Napier's analogies* (the word *analogy* being employed in its archaic sense of "proportion"):

$$\frac{\sin (A - B)/2}{\sin (A + B)/2} = \frac{\tan (a - b)/2}{\tan c/2} \, ,$$

$$\frac{\cos (A - B)/2}{\cos (A + B)/2} = \frac{\tan (a + b)/2}{\tan c/2} \, ,$$

$$\frac{\sin (a - b)/2}{\sin (a + b)/2} = \frac{\tan (A - B)/2}{\cot C/2} \, ,$$

$$\frac{\cos (a - b)/2}{\cos (a + b)/2} = \frac{\tan (A + B)/2}{\cot C/2} \, .$$

These formulas, which are analogous to the law of tangents in plane trigonometry, may be used to solve oblique spherical triangles for which the given parts are two sides and the included angle or two angles and the included side.

(a) Find A, C, b for a spherical triangle in which $a = 125°\ 38'$, $C = 73°\ 24'$, $B = 102°\ 16'$.

(b) Find A, B, c for a spherical triangle in which $a = 93°\ 8'$, $b = 46°\ 4'$, $C = 71°\ 6'$.

17.6. The difficulty that was so widely experienced in the multiplication of large numbers led to mechanical ways of carrying out the process. Very celebrated in its time was Napier's invention,

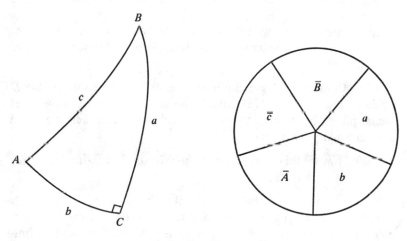

FIG. 57

known as *Napier's rods,* or *Napier's bones,* and described by the inventor in his work *Rabdologiae,* published in 1617. In principle the invention is the same as the Hindu lattice, or grating, method which is described in Exercise 13.9, except that in the invention the process is carried out with the aid of rectangular strips of bone, metal, wood, or cardboard, prepared beforehand. For each of the ten digits one should have a set of strips, like the one shown to the left in Figure 58 for the digit 6, bearing the various multiples of that digit. To illustrate the use of these strips in multiplication, let us select the example chosen by Napier in the *Rabdologiae,* the multiplication of 1615 by 365. Put strips headed with 1, 6, 1, 5 side by side as shown to the right in Figure 58. The results of multiplying 1615 by the 5, the 6, and the 3 of 365 are then easily read off as 8075, 9690, and 4845, some simple diagonal additions of two digits being necessary to obtain these results. The final product is then found by an addition, as illustrated in the figure.

(a) Make a set of Napier's rods and perform some multiplications.

(b) Explain how Napier's rods may be used to perform divisions.

17.7. (a) Construct, with the aid of tables, a logarithmic scale, to be designated as the D scale, about 10 inches long. Use the scale, along with a pair of dividers, to perform some multiplications and divisions.

(b) Construct two logarithmic scales, to be called C and D scales, of the same size. By sliding C along D perform some multiplications and divisions.

17.8. Construct a logarithmic scale exactly half as long as the D scale in Exercise 17.7. (a), and designate by A two of these short scales placed in tandem. Show how the A and D scales may be used for extracting square roots.

17.9. How would one design a scale to be used with the D scale for extracting cube roots?

17.10. Construct a scale exactly like the C and D scales, only running in the reverse direction, and call it the CI (*C inverted*) scale. Show how the CI and D scales may be used for performing multiplications. What is the advantage of the CI and D scales over the C and D scales for this purpose?

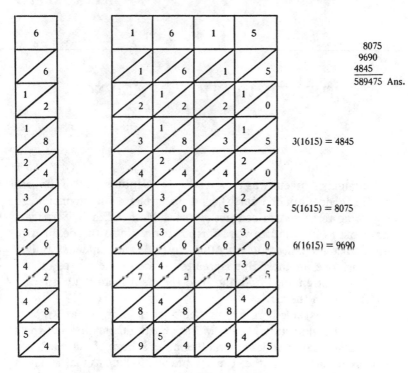

8075
9690
4845
─────
589475 Ans.

3(1615) = 4845

5(1615) = 8075

6(1615) = 9690

FIG. 58

Further Reading

COOLIDGE, J. L., *The Mathematics of Great Amateurs*. New York: Oxford University Press, 1949.

HOBSON, E. W., *John Napier and the Invention of Logarithms*. New York: Cambridge University Press, 1914.

KNOTT, C. G., *Napier Tercentenary Memorial Volume*. London: Longmans, Green, 1915.

THE STIMULATION OF SCIENCE

The mighty Antaeus was the giant son of Neptune (god of the sea) and Ge (goddess of the earth), and his strength was invincible so long as he remained in contact with his Mother Earth. Strangers who came to his country were forced to wrestle to the death with him, and so it chanced one day that Hercules and Antaeus came to grips with one another. But Hercules, aware of the source of Antaeus' great strength, lifted and held the giant from the earth and crushed him in the air.

There is a parable here for mathematicians. For just as Antaeus was born of and nurtured by his Mother Earth, history has shown us that all significant and lasting mathematics is born of and nurtured by the real world. As in the case of Antaeus, so long as mathematics maintains its contact with the real world, it will remain powerful. But should it be lifted too long from the solid ground of its birth into the filmy air of pure abstraction, it runs the risk of weakening. It must of necessity return, at least occasionally, to the real world for renewed strength.

Such a rejuvenation of mathematics occurred in the seventeenth century, following discoveries made by two eminent mathematician-scientists—Galileo Galilei (1564-1642) and Johann Kepler (1554-1630). Galileo, through a sequence of experiments started before his twenty-fifth birthday, discovered a number of basic facts concerning the motion of bodies in the earth's gravitational field, and Kepler, by 1619, had induced all three of his famous laws of planetary motion. These achievements proved to be so influential on the development of so much of subsequent mathematics that they must be ranked as two of the GREAT MOMENTS IN MATHEMATICS. Galileo's discoveries led to the creation of the modern science of dynamics and Kepler's to the creation of modern celestial mechanics; and each of these studies, in turn, required, for their development, the creation of a new mathematical tool—the calculus—capable of dealing with change, flux, and motion.

A new type of mathematics came into being. In contrasting the older and the newer mathematics, the older appears passive and static while the newer appears vigorous and dynamic, so that the older mathematics compares to the still-picture stage of photography while the new mathematics compares to the moving-picture stage. Again, the older mathematics is to the newer as anatomy is to physiology, wherein the former studies the dead body and the latter the living body. Once more, the older mathematics concerned itself with the fixed and the finite while the newer mathematics embraces the changing and the infinite.

Galileo was born in Pisa in 1564 as the son of an impoverished Florentine nobleman. After a disinterested start as a medical student, Galileo obtained parental permission to change his studies to science and mathematics, fields in which he possessed a strong natural talent. While still a medical student at the University of Pisa, he made his historically famous observation that the great pendulous lamp in the cathedral there oscillated to and fro with a period independent of the size of the arc of oscillation.* Later he showed that the period of a pendulum is also independent of the weight of the pendulum's bob. When he was 25, he accepted an appointment as professor of mathematics at the University of Pisa. It was during this appointment that he is alleged to have performed experiments from the leaning tower of Pisa, showing that, contrary to the teaching of Aristotle, heavy bodies do not fall faster than light ones. By rolling balls down inclined planes, he arrived at the law that the distance a body falls is proportional to the square of the time of falling, in accordance with the now familiar formula $s = (\frac{1}{2})gt^2$.

Unpleasant local controversies caused Galileo to resign his chair at Pisa in 1591, and the following year he accepted a professorship in mathematics at the University of Padua, where there reigned an atmosphere more friendly to scientific pursuits. Here at Padua, for

---

*This is only approximately true, the approximation being very close in the case of small amplitudes of oscillation.

nearly 18 years, Galileo continued his experiments and his teaching, achieving a widespread fame. While at Padua, he heard of the discovery, in about 1607, of the telescope by the Dutch lens-grinder Johann Lippersheim, and he set about making instruments of his own, producing a telescope with a magnifying power of more than 30 diameters. With this telescope he observed sunspots (contradicting Aristotle's assertion that the sun is without blemish), saw mountains on the moon, and noticed the phases of Venus, Saturn's rings, and the four bright satellites of Jupiter (all three of these lending credence to the Copernican theory of the solar system). Galileo's discoveries roused the opposition of the Church, and finally, in the year 1633, he was summoned to appear before the Inquisition, and there forced to recant his scientific findings. Not many years later the great scientist became blind. He died, a prisoner in his own home, in 1642, the year Isaac Newton was born.

It is to Galileo that we owe the modern spirit of science as a harmony between experiment and theory. He not only founded the mechanics of freely falling bodies, but he laid the foundations of dynamics in general, foundations upon which Newton was later able to build the science mathematically. Galileo was the first to realize the parabolic nature of the path of a projectile in a vacuum, and he speculated on the laws of momentum. He invented the first modern-type microscope. As early as 1597 he perfected the *sector compasses*, a simple instrument that enjoyed wide popularity for more than two centuries.

Galileo wrote, in Italian, two famous treatises, one on astronomy and one on physics. Giving them their English titles, they are *The Two Chief Systems* (1632), concerning the relative merits of the Ptolemaic and Copernican views of the universe, and *The Two New Sciences* (1638), devoted to dynamics and the strength of materials. Each work is in the form of a dialogue among three men: Salviati (an informed science scholar), Sagredo (an intelligent layman), and Simplicio (an orthodox Aristotelian). It was the first book that directly led to Galileo's trial and confinement; the second book, published in Leyden, was written during Galileo's unhappy years of forced detention. In these treatises can be found recognition of certain properties of the infinitely large and the infinitely small, such as the idea of the equivalence of infinite classes that later became fundamental in Cantor's nineteenth-century theory of sets and transfinite numbers.

It would seem that Galileo may have been jealous of his famous contemporary, Johann Kepler, for although Kepler had announced all three of his important laws of planetary motion by 1619, these laws were completely ignored by Galileo.

Johann Kepler was born near Stuttgart, Germany, in 1571, and commenced his studies at the University of Tübingen with the intention of becoming a Lutheran minister. Like Galileo, he found his first choice of an occupation far less congenial than his deep interest in science, particularly astronomy, and he accordingly changed his plans. In 1594, when in his early twenties, he accepted a lectureship at the University of Grätz in Austria. Five years later he became assistant to the famous Danish-Swedish astronomer Tycho Brahe, who had moved to Prague to serve as the court astronomer to Kaiser Rudolph II. Shortly after, in 1601, Brahe suddenly died, and Kepler inherited both his master's position and his vast collection of very accurate data on the positions of the planets as they moved about the sky. With amazing perseverance, Kepler set out to find, from Brahe's enormous mass of observational data, just how the planets move in space.

It has often been remarked that almost any problem can be solved if one but continuously worries over it and works at it a sufficiently long time. Somewhat as Thomas Edison said of invention being 1 percent inspiration and 99 percent perspiration, problem solving is 1 percent imagination and 99 percent perseverance. Perhaps nowhere in the history of science is this more clearly demonstrated than in Kepler's incredible pertinacity in solving the problem of the motion of the planets about the sun. Thoroughly convinced of the Copernican theory that the planets revolve in orbits about the central sun, Kepler strenuously sought to determine the nature and position of those orbits and the manner in which the planets travel in their orbits. With Brahe's great set of observational recordings at hand, the problem became this: to obtain a pattern of motion of the planets that would exactly agree with Brahe's observations. So dependable were Brahe's recordings that any solution that should differ from Brahe's observed positions by even as little as a quarter of the moon's apparent diameter must be discarded as incorrect. Kepler had, then, first to guess with his *imagination* some plausible solution, and then with painful *perseverance* to endure the mountain of tedious calculation needed to confirm or reject his guess. He made hundreds of fruitless attempts and performed reams and reams of calculations, laboring with undiminished zeal and patience for many years. Finally he solved his problem, in the form of his three famous laws of planetary motion, the first two found in 1609 and the third one ten years later in 1619:

I. *The planets move about the sun in elliptical orbits with the sun at one focus.*

II. *The radius vector joining a planet to the sun sweeps over equal areas in equal intervals of time.*

III. *The square of the time of one complete revolution of a planet about its orbit is proportional to the cube of the orbit's semimajor axis.*

The empirical discovery of these laws from Brahe's mass of data constitutes one of the most remarkable inductions ever made in science. With justifiable pride, Kepler prefaced his *Harmony of the Worlds* of 1619 with the following poetic outburst:

> I am writing a book for my contemporaries or—it does not matter—for posterity. It may be that my book will wait for a hundred years for a reader. Has not God waited for 6000 years for an observer?

Kepler's laws of planetary motion are landmarks in the history of astronomy and mathematics, for in the effort to justify them Isaac Newton was led to create modern celestial mechanics. It is very interesting that 1800 years after the Greeks had developed the properties of the conic sections there should occur such an illuminating practical application of them. One never knows when a piece of pure mathematics may receive an unexpected application.

In order to compute the areas involved in his second law, Kepler had to resort to a crude form of the integral calculus, making him one of the precursors of that calculus. He also, in his *Stereometria doliorum vinorum* (Solid Geometry of Wine Barrels, 1615), applied crude integration procedures to the finding of the volumes of 93 different solids obtained by revolving arcs of conic sections about axes in their planes. Among these solids were the torus and two that he called *the apple* and *the lemon,* these latter being obtained by revolving a major and a minor arc, respectively, of a circle about the arc's chord as an axis. Kepler's interest in these matters arose when he observed some of the poor methods in use by the wine gaugers of the time. It is quite possible that Cavalieri was influenced by this work of Kepler when he later carried the refinement of the integral calculus a stage further with his *method of indivisibles,* to be discussed in our next lecture.

Kepler also made notable contributions to the subject of polyhedra. He seems to have been the first to recognize an *antiprism* (obtained from a prism by rotating the top base in its own plane so as to make its vertices correspond to the sides of the lower base, and then joining in zigzag fashion the vertices of the two bases). He also discovered the cuboctahedron, rhombic dodecahedron, and rhombic triakontahedron. The second of these polyhedra occurs in nature as a garnet crystal. Of the four possible star-polyhedra, two were discovered by Kepler and the other two in 1809 by Louis Poinsot (1777–1859), a pioneer worker in geometrical mechanics. The Kepler-Poinsot star-polyhedra are space analogues of the regular star-polygons in the plane. Kepler was also a pioneer investigator in the problem of covering the plane with regular polygons (not necessarily all alike).

It was Kepler who introduced the word *focus* (Latin for "hearthside") into the geometry of conic sections. He approximated the perimeter of an ellipse of semiaxes a and b by the formula $\pi(a + b)$. He also laid down a so-called *principle of continuity* which postulates the existence in a plane of certain ideal points and an ideal line, having many of the properties of ordinary points and lines, lying at infinity. Thus he explained that a straight line can be considered as closed at infinity, that two parallel lines should be regarded as intersecting at infinity, and that a parabola may be regarded as the limiting case of either an ellipse or a hyperbola in which one of the foci has retreated to infinity. These ideas were extended by later geometers.

Kepler was a confirmed Pythagorean, with the result that his work is often a blend of the fancifully mystical and the carefully scientific. It is sad that his personal life was made almost unendurable by a multiplicity of worldly misfortunes. An infection from smallpox when he was but four years old left his eyesight much impaired. In addition to his general lifelong weakness, he spent a joyless youth; his marriage was a constant source of unhappiness; his favorite child died of smallpox; his wife went mad and died; he was expelled from his lectureship at the University of Grätz when the city fell to the Catholics; his mother was charged and imprisoned for witchcraft, and for almost a year he desperately tried to save her from the torture chamber; he himself very narrowly escaped condemnation for heterodoxy; and his stipend was always in arrears. One report says that his second marriage was even less fortunate than his first, although he took the precaution to analyze carefully the merits and demerits of eleven girls before choosing the wrong one. He was forced to augment his income by casting horoscopes, and he died of a fever in 1630 at the age of 59 while on a journey to try to collect some of his long overdue salary.

Exercises

18.1. Assuming that all bodies fall with the same constant acceleration g, Galileo showed that the distance d a body falls is proportional to the square of the time t of falling. Establish the following stages of Galileo's argument.

(a) If v is the velocity at the end of time t, then $v = gt$.

(b) If v and t refer to one falling body and V and T to a second falling body, then $v/V = t/T$, whence the right triangle having legs of numerical lengths v and t is similar to the right triangle having legs of numerical lengths V and T.

(c) Since the increase in velocity is uniform, the average velocity of fall is $(\frac{1}{2})v$, whence $d = (\frac{1}{2})vt$ = area of right triangle with legs v and t.

(d) $d/D = t^2/T^2$. Show also that $d = (\frac{1}{2})gt^2$.

Galileo illustrated the truth of this final law by observing the times of descent of balls rolling down inclined planes.

18.2. A *sector compasses* consists of two arms fastened together at one end by a pivot joint, as shown in Figure 59. On each arm there is

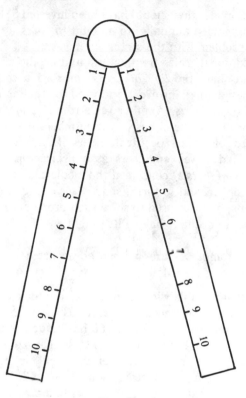

FIG. 59

a simple scale radiating from the pivot and having the zero of the scale at the pivot.

(a) Show how the sector compasses may be used to divide a given line segment into five equal parts.

(b) Show how the sector compasses may be used to change the scale of a drawing.

18.3. (a) Show how the sector compasses (see Exercise 18.2) may be used to find the fourth proportional x to three given quantities a, b, c (that is, to find x where $a : b = c : x$), and thus applied to a problem in foreign exchange.

(b) Galileo illustrated the use of the sector compasses by finding the amount of money that should have been invested 5 years ago at 6 percent, compounded annually, to amount to 150 scudi today. Try to solve this problem with the sector compasses.

Among additional scales frequently found on the arms of sector compasses was one (the *line of areas*) marked according to the squares of the numbers involved, and used for finding squares and square roots of numbers. Another scale (the *line of volumes*) was marked according to the cubes of the numbers involved. Another gave the chords of arcs of specific numbers of degrees for a circle of unit radius, and served engineers as a protractor. Still another (called the *line of metals*) contained the medieval symbols for gold, silver, iron, copper, and so forth, spaced according to the densities of these metals, and was used to solve such problems as finding the diameter of an iron sphere having its weight equal to that of a given copper sphere.

The sector compasses are neither as accurate nor as easy to manipulate as the slide rule.

18.4. Explain the following geometrical paradox considered by Galileo in his *The Two New Sciences* of 1638.

Suppose the large circle of Figure 60 has made one revolution in rolling along the straight line from A to B, so that AB is equal to the circumference of the large circle. Then the small circle, fixed to the large one, has also made one revolution, so that CD is equal to the circumference of the small circle. It follows that *the two circles have equal circumferences.*

This paradox had been earlier described by Aristotle and is therefore sometimes referred to as *Aristotle's wheel.*

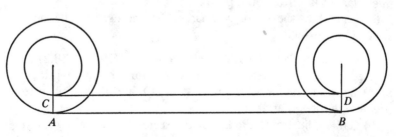

Fig. 60

18.5. Explain the remark in Galileo's *The Two New Sciences* that "neither is the number of squares less than the totality of all numbers, nor the latter greater than the former."

18.6. (a) Where is a planet in its orbit when its speed is greatest?

(b) Check, approximately, Kepler's third law using the following modern figures. (A.U. is an abbreviation for *astronomical unit,* the length of the semimajor axis of the earth's orbit.)

Planet	Time in years	Semimajor axis
Mercury	0.241	0.387 A.U.
Venus	0.615	0.723 A.U.
Earth	1.000	1.000 A.U.
Mars	1.881	1.524 A.U.
Jupiter	11.862	5.202 A.U.
Saturn	29.457	9.539 A.U.

(c) What would be the period of a planet having a semimajor axis of 100 A.U.?

(d) What would be the semimajor axis of a planet having a period of 125 years?

18.7. (a) Two hypothetical planets are moving about the sun in elliptical orbits having equal major axes. The minor axis of one, however, is half that of the other. How do the periods of the two planets compare?

(b) The moon revolves about the earth in 27.3 days in an elliptical orbit whose semimajor axis is 60 times the earth's radius. What would be the period of a satellite revolving close to the earth's surface?

18.8. A very interesting problem of tiling is to fill the plane with congruent regular polygons. Let n be the number of sides of each polygon.

(a) If we do not permit a vertex of one polygon to lie on a side of another, show that the number of polygons at each vertex is given by

$$2 + 4/(n - 2),$$

and hence that we must have $n = 3$, 4, or 6. Construct illustrative tilings.

(b) If we insist that a vertex of one polygon lie on a side of another, show that the number of polygons clustered at such a vertex is given by

$$1 + 2/(n - 2),$$

whence we must have $n = 3$ or 4. Construct illustrative tilings.

18.9. Construct tilings of the plane containing

(a) two sizes of equilateral triangles, the larger having a side twice that of the smaller, and such that sides of triangles of the same size do not overlap,

(b) two sizes of squares, the larger having a side twice that of the smaller, and such that sides of smaller squares do not overlap,

(c) congruent equilateral triangles and congruent regular dodecagons,

(d) congruent equilateral triangles and congruent regular hexagons,

(e) congruent squares and congruent regular octagons.

(f) Suppose we have a tiling of the plane composed of regular polygons of three different kinds at each vertex. If the three kinds of polygons have p, q, r sides, respectively, show that

$$\frac{1}{p} + \frac{1}{q} + \frac{1}{r} = \frac{1}{2}.$$

One integral solution of this equation is $p = 4$, $q = 6$, $r = 12$. Construct a tiling of the plane of the type under consideration and composed of congruent squares, congruent regular hexagons, and congruent regular dodecagons.

18.10. (a) Kepler divided a circle of radius r and circumference C into a large number of very thin sectors. By regarding each sector as a thin isosceles triangle of altitude r and base equal to the arc of the sector, he heuristically arrived at the formula $A = (\frac{1}{2})rC$ for the area of a circle. Show how this was done.

(b) Use Kepler's type of reasoning to obtain the volume V of a sphere of radius r and surface area S as $V = (\frac{1}{3})rS$.

(c) Use Kepler's type of reasoning to obtain the area A of the ellipse of semimajor axis a and semiminor axis b as $A = \pi ab$.

Further Reading

BRASCH, F. F., ed., *Johann Kepler, 1571–1630. A Tercentenary Commemoration of His Life and Work.* Baltimore: Williams and Wilkins, 1931.

CASPAR, MAX, *Kepler,* tr. by C. Doris Hellman. New York: Abelard-Schuman, 1959.

FAHIE, J. J., *Galileo, His Life and Work.* London: John Murray, 1903.

PÓLYA, GEORGE, *Mathematical Methods in Science* (New Mathematical Library, No. 26). Washington, D.C.: The Mathematical Association of America, 1977.

SLICING IT THIN

In the fourteenth century, the Blessed John Colombini of Siena founded a religious group known as the *Jesuats*, which was in no way related to the *Jesuits*. The order was approved by Pope Urban V in 1367. The original work of the order was the care of those stricken by the Black Death, which raged over Europe at the time, and the burial of the fatally smitten. With the passage of time the Jesuat order diminished, and in 1606 an attempt at a revival was made. But certain abuses later crept into the order, with the result that the group now no longer exists. It seems that the manufacture and sale of distilled liquors, apparently in a manner unacceptable by Canon Law, along with a growing scarcity of members, led to the order's suppression by Pope Clement IX in 1668.

In 1613, only a few years after the attempted revival of the Jesuats, a young fifteen-year-old Italian boy named Bonaventura Cavalieri was accepted as a member of the order, and then spent the rest of his life in its service. It is because of this, and because of the ultimate vanishing of the order and the natural confusion between Jesuat and Jesuit, that so many major encyclopedias, histories, and source books erroneously state that Cavalieri was a Jesuit, instead of a Jesuat, furnishing an excellent example of written histories containing a hidden perpetuated error. It is all too easy for some historian to record an erroneous and undocumented statement, and then for subsequent historians, leaning on the earlier work, to repeat the falsehood. Many such erroneous statements have been widely perpetuated over considerable periods of time.

Bonaventura Cavalieri was born in Milan, Italy, in 1598, studied under Galileo, and served as a professor of mathematics at the University of Bologna from 1629 until his death in 1647 at the age of forty-nine. Cavalieri was one of the most influential mathematicians of his time, and the author of a number of works on trigonometry, geometry, optics, astronomy, and astrology. He was among the first to recognize the great value of logarithms and was largely responsible for their early introduction into Italy. But his greatest contribution to mathematics was a treatise, *Geometria indivisibilibus,* published in its first form in 1635, devoted to the pre-calculus *method of indivisibles*—a method that can, like so many things in more modern mathematics, be traced back to the early Greeks, in this case to Democritus (ca. 410 B.C.) and Archimedes (ca. 287–212 B.C.). It is quite likely that it was the attempts at integration made by Kepler that directly motivated Cavalieri. At any rate, the publishing of Cavalieri's *Geometria indivisibilibus* in 1635 marks a GREAT MOMENT IN MATHEMATICS.

Cavalieri's treatise on the method of indivisibles is voluble and not clearly written, and it is not easy to learn from it precisely what Cavalieri meant by an "indivisible." It seems that an indivisible of a given planar piece is a chord of the piece, and a planar piece can be considered as made up of an infinite parallel set of such indivisibles. Similarly, it seems that an indivisible of a given solid is a planar section of that solid, and a solid can be considered as made up of an infinite parallel set of this kind of indivisible. Now, Cavalieri argued, if we slide each member of a parallel set of indivisibles of some planar piece along its own axis, so that the endpoints of the indivisibles still trace a continuous boundary, then the area of the new planar piece so formed is the same as that of the original planar piece, inasmuch as the two pieces are made up of the same indivisibles. A similar sliding of the members of a parallel set of indivisibles of a given solid will yield another solid having the same volume as the original one. (This last result can be strikingly illustrated by taking a vertical stack of cards and then pushing the sides of the stack into curved surfaces; the volume of the disarranged stack is the same as that of the original stack.) These results, slightly generalized, give the so-called *Cavalieri principles:*

1. *If two planar pieces are included between a pair of parallel lines, and if the lengths of the two segments cut by them on any line parallel to the including lines are always in a given ratio, then the areas of the two planar pieces are also in this ratio.*

2. *If two solids are included between a pair of parallel planes, and if the areas of the two sections cut by them on any plane parallel to the including planes are always in a given ratio, then the volumes of the two solids are also in this ratio.*

Cavalieri's principles constitute a valuable tool in the computation of areas and volumes, and their intuitive bases can easily be made rigorous with the modern integral calculus. Accepting these principles as intuitively apparent, one can solve many problems in mensuration that normally require the more advanced techniques of the calculus.

Let us illustrate the use of Cavalieri's principles, first employing the planar case to find the area of an ellipse of semiaxes a and b, and then the solid case to find the volume of a sphere of radius r.

Consider the ellipse and circle

$$x^2/a^2 + y^2/b^2 = 1, \quad a > b, \quad \text{and} \quad x^2 + y^2 = a^2,$$

plotted on the same rectangular coordinate frame of reference, as shown in Figure 61. Solving each of the equations above for y, we find, respectively,

$$y = (b/a)(a^2 - x^2)^{1/2}, \quad y = (a^2 - x^2)^{1/2}.$$

It follows that corresponding ordinates of the ellipse and the circle are in the ratio b/a. It then follows that corresponding vertical chords of the ellipse and the circle are also in this ratio, and hence, by Cavalieri's first principle, so are the areas of the ellipse and the circle. We conclude that

$$\text{area of ellipse} = (b/a)(\text{area of circle})$$
$$= (b/a)(\pi a^2) = \pi ab.$$

This is basically the procedure Kepler employed in finding the area of an ellipse of semiaxes a and b.

Now let us find the familiar formula for the volume of a sphere of radius r. In Figure 62 we have a hemisphere of radius r on the left, and on the right a circular cylinder of radius r and altitude r with a cone removed whose base is the upper base of the cylinder and whose vertex is the center of the lower base of the cylinder. The hemisphere and the gouged-out cylinder are resting on a common plane. We now cut both solids by a plane parallel to the base plane and at

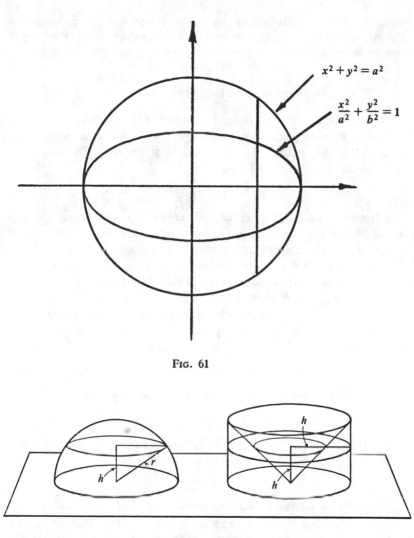

$$x^2 + y^2 = a^2$$

$$\frac{x^2}{a^2} + \frac{y^2}{b^2} = 1$$

FIG. 61

FIG. 62

distance h above it. This plane cuts the one solid in a circular section and the other in an annular, or ring-shaped, section. By elementary geometry we easily show that each of the two sections has an area equal to $\pi(r^2 - h^2)$. It follows, by Cavalieri's second principle, that the two solids have equal volumes. Therefore the volume V of the sphere is given by

$$V = 2 \text{ (volume of cylinder} - \text{volume of cone)}$$
$$= 2 (\pi r^3 - \pi r^3/3) = 4\pi r^3/3.$$

As another illustration of the planar case of Cavalieri's principles, consider the area pictured in the left of Figure 63. One easily obtains, for a comparison area, that pictured in the right of Figure 63. It follows that the required area is

$$A = (\pi/2 + 1)r^2.$$

For a second illustration of the solid case of Cavalieri's principles, let us find the volume of the *spherical ring* obtained by removing from a solid sphere of radius r a cylindrical boring of radius a coaxial with the polar axis of the sphere (pictured in the left of Figure 64). For a comparison solid use a sphere of diameter equal to the altitude of the spherical ring (pictured in the right of Figure 64). Now cut the two solids by a horizontal plane at distance h from the centers of the two solids. In the spherical ring we get an annular section of area

$$\pi(r^2 - h^2) - \pi a^2 = \pi(r^2 - a^2 - h^2).$$

In the sphere we get a circular section of area

$$\pi(k^2 - h^2) = \pi(r^2 - a^2 - h^2).$$

It follows that the volume V of the spherical ring is the same as the volume of the sphere of radius k. That is

$$V = 4\pi k^3/3.$$

It interestingly follows that *all spherical rings of the same altitude have the same volume, irrespective of the radii of the rings.*

The assumption and then consistent use of Cavalieri's second principle can greatly simplify the derivation of many of the volume formulas encountered in a beginning treatment of solid geometry. This procedure has been adopted by a number of textbook writers, and has been advocated on pedagogical grounds. For example, in deriving the familiar formula for the volume of a tetrahedron ($V =$

Bh/3), the sticky part is first to show that any two tetrahedra having equivalent bases and equal altitudes on those bases have equal volumes. The inherent difficulty here is reflected in all treatments of solid geometry from Euclid's *Elements* on. With Cavalieri's second principle, however, the difficulty simply melts away.

Cavalieri's hazy conception of indivisibles, as sort of atomic parts of a figure, led to much discussion and serious criticism by some students of the subject, particularly by the Swiss goldsmith and mathematician Paul Guldin (1577–1642). Cavalieri recast his treatment in the vain hope of meeting these objections. The French geometer and physicist Gilles Persone de Roberval (1602–1675) ably

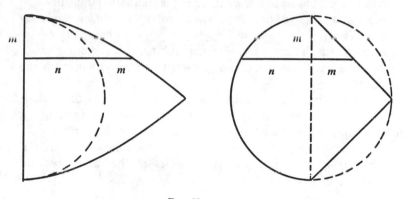

FIG. 63

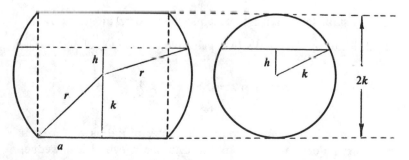

FIG. 64

employed the method and claimed to be an independent inventor of it. The method, or some process very like it, was effectively used by Evangelista Torricelli (1608–1647), Pierre de Fermat (1601?–1665), Blaise Pascal (1623–1662), Grégoire de Saint-Vincent (1584–1667), Isaac Barrow (1630–1677), and others. In the course of the work of these men, results were reached which are equivalent to performing such integrations as

$$\int x^n \, dx, \qquad \int \sin x \, dx, \qquad \int \sin^2 x \, dx, \qquad \int x \sin x \, dx.$$

Two planar pieces that can be placed so that they cut off equal segments on each member of a family of parallel lines, or two solids that can be placed so that they intercept equiareal sections on each member of a family of parallel planes, are said to be *Cavalieri congruent*. Two figures that are Cavalieri congruent have, of course, equal areas (in the one case) or equal volumes (in the other case). Among the curiosities concerning this type of congruence are the following:

1. Though there cannot exist a polygon to which a given circle is Cavalieri congruent, there does exist a polyhedron (actually a tetrahedron) to which a given sphere is Cavalieri congruent.

2. Though there exist tetrahedra of the same volume that are not Cavalieri congruent, any pair of triangles of the same area are Cavalieri congruent.

Exercises

19.1. Establish Cavalieri's principles by modern integration.

19.2. Find, by Cavalieri's first principle, the area enclosed by the curve

$$b^2 y^2 = (b + x)^2 (a^2 - x^2),$$

where $b \geq a > 0$.

19.3. An oblique plane through the center of the base of a right circular cylinder cuts off from the cylinder a cylindrical wedge, called a *hoof*. Find, by Cavalieri's second principle, the volume of a hoof in terms of the radius r of the associated cylinder and the altitude h of the hoof.

19.4. Show that there cannot exist a polygon to which a given circle is Cavalieri congruent.

19.5. Find a polyhedron to which a given sphere is Cavalieri congruent.

19.6. Find, by Cavalieri's second principle, the volume of a *torus*, or *anchor ring*, formed by revolving a circle of radius r about a line in the plane of the circle at distance $c \geq r$ from the center of the circle.

19.7. (a) Show that any triangular prism can be dissected into three triangular pyramids having, in pairs, equivalent bases and equal altitudes.

(b) Show, by Cavalieri's second principle, that two triangular pyramids having equivalent bases and equal altitudes have equal volumes.

(c) Show that the volume of a triangular pyramid is equal to one third the product of the area of the base of the pyramid and the altitude of the pyramid.

19.8. A *generalized prismoid* is any solid having two parallel base planes and having the areas of its sections parallel to the bases given by a quadratic function of their distances from one base.

(a) Show that the volumes of a *prism,* a *wedge* (a right triangular prism turned so as to rest on one of its lateral faces as a base), and a *pyramid* are given by the *prismoidal formula*

$$V = h(U + 4M + L)/6,$$

where h is the altitude, and U, L, and M are the areas of the upper and lower bases and midsection, respectively.

(b) Show, by Cavalieri's second principle, that the volume of a generalized prismoid is given by the prismoidal formula.

(c) Establish part (b) by integral calculus.

19.9. Show that (a) a *sphere,* (b) an *ellipsoid,* (c) a *hoof* (see Exercise 19.3), and (d) a *Steinmetz solid* (the solid common to two right circular cylinders of equal radii and having their axes intersecting perpendicularly) are all examples of a generalized prismoid, and thus find expressions for their volumes.

19.10. Two solids, included between a pair of parallel planes, are such that the perimeters of the two sections cut by them on any plane parallel to the including planes are always equal in length. Does it follow that the lateral areas of the two solids are equal?

Further Reading

BOYER, C. B., *The Concepts of the Calculus*. New York: Dover, 1959.

LECTURE **20**

THE TRANSFORM-SOLVE-INVERT TECHNIQUE

One of the most effective methods employed by mathematicians is that known as the *transform-solve-invert technique*. The essence of the idea is this. To solve a difficult problem, *transform* it, by some simplifying procedure, into an easier equivalent problem, *solve* the easier problem, and then *invert* the simplifying procedure to obtain a solution of the original problem. Let us illustrate the idea with some examples.

Suppose someone should ask us a question in French and that we are not overly proficient in that language. We would first *transform* the question into an equivalent one in English, a language in which we are much more facile. We would then *solve* the matter in English by phrasing an English answer to the question. Finally, we would *invert* our English answer into French, thus solving the original problem.

Again, suppose we are asked to find the Roman numeral representing the product of the two given Roman numerals LXIII and XXIV. We would *transform* the two given Roman numerals into the corresponding Hindu-Arabic numerals, 63 and 24. We would then *solve* the related problem in the Hindu-Arabic notation by means of the familiar multiplication algorithm to obtain the product 1512. Finally, we would *invert* this result back into Roman notation, obtaining MDXII as the answer to the original problem. By an appropriate transformation, a difficult problem has been converted into an easy problem.

As a more sophisticated problem, suppose we are asked to show that the equation

$$x^7 - 2x^5 + 10x^2 - 1 = 0$$

215

has no root greater than 1. By the substitution $x = y + 1$ we *transform* the given equation into

$$y^7 + 7y^6 + 19y^5 + 25y^4 + 15y^3 + 11y^2 + 17y + 8 = 0.$$

Since the roots of the new equation are equal to those of the original equation diminished by 1 (inasmuch as $y = x - 1$), we must show that the new equation has no root greater than 0. We *solve* this equivalent problem simply by observing that all the coefficients in the new equation are positive, whence y cannot be positive and yet yield a zero sum. Finally, we *invert* the transformation to obtain the desired result.

As a concluding example, suppose we are asked (see Figure 65) to prove that the area of the curvilinear triangle bounded by three congruent and similarly oriented ellipses that touch each other externally in pairs is independent of the relative positions of the three ellipses. Now it is known that by a suitable orthogonal projection we may project any ellipse into a circle whose radius is the semiminor axis of the ellipse. Let us *transform* the figure of our problem by the orthogonal projection that will carry our three tangent congruent ellipses into three tangent congruent circles (see Figure 66). Since it is known that under an orthogonal pro-

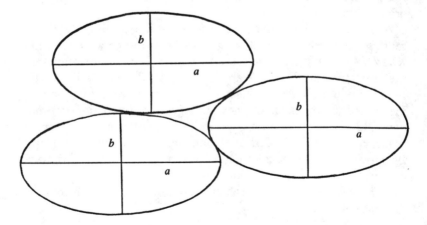

Fig. 65

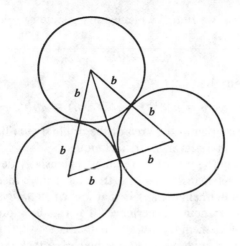

FIG. 66

jection equal areas project into equal areas, all we must show is that in the projected figure the area of the curvilinear triangle is independent of the relative positions of the three congruent circles. We easily *solve* this, as it is obvious. Finally, we *invert* the orthogonal projection to obtain a solution of the original problem.

The procedure above is useful not only in solving problems but also in discovering new facts. We *transform* a given mathematical setup into a new one. By studying the new setup we *discover* some property of it. We then *invert* the transformation to obtain a property of the original setup. This technique is much used by mathematicians and may be called the *transform-discover-invert technique*.

Consider, for example, our last example above—the one involving an orthogonal projection p. Now it is known that under p an area A projects into an area $A' = A \cos \theta$, where θ is the angle between the plane of the original figure and the plane of the projected figure. In the case of our example it is easy to show that $\cos \theta = b/a$, where a and b are the semiaxes of our ellipses. Letting A and A' denote the areas of the curvilinear triangle before and after projection, we find, by elementary geometry (see Figure 66), that

$$A' = b^2(\sqrt{3} - \pi/2).$$

It follows, then, that

$$A = A' \sec \theta = b^2(\sqrt{3} - \pi/2)(a/b) = ab(\sqrt{3} - \pi/2).$$

We have cleverly found the area of the original curvilinear triangle by the transform-discover-invert technique.

Far and away the grandest, the most extensively developed, and the most productive instance of the transform-solve-invert technique yet devised by mathematicians is that known as *analytic geometry*. There are few academic experiences that can be more thrilling to the student of elementary college mathematics than his introduction to this new and powerful method of attacking geometrical problems—for analytic geometry is a *method,* rather than a *branch,* of geometry.

The essence of the idea of analytic geometry as applied to the plane, it will be recalled, is the establishment of a correspondence between ordered pairs of real numbers and points in the plane, thereby making possible a correspondence between curves in the plane and equations in two variables, so that for each curve in the plane there is a definite equation $f(x,y) = 0$, and for each such equation there is a definite curve, or set of points, in the plane. A correspondence is similarly established between the algebraic and analytic properties of the equation $f(x,y) = 0$ and the geometric properties of the associated curve. The task of proving a theorem in geometry is cleverly shifted to that of proving a corresponding theorem in algebra or analysis. The method goes deeper than this, for a noted algebraic or analytical result can lead to the discovery of a new and unsuspected geometrical result. Analytic geometry thus proves to be a highly productive method, both for solving problems and for discovering new results in geometry.

There is no unanimity of opinion among historians of mathematics concerning who invented analytic geometry, nor even concerning what age should be credited with the invention. Much of this difference of opinion stems from a lack of agreement regarding just what constitutes analytic geometry. There are those who, favoring antiquity as the era of the invention, point out the well-known fact that the concept of fixing the position of a point by means of suitable coordinates was employed in the ancient world by the Egyptians and the Romans in surveying, and by the Greeks in map making. And, if analytic geometry implies not only the use of coordinates but also the geometric interpretation of relations among coordinates, then particularly strong in the favor of the Greeks is the fact that Apollonius derived the bulk of his geometry of the conic sections from the geometrical equivalents of certain Cartesian equations of these curves, an idea that seems to have originated with Menaechmus about 350 B.C. Others claim that the invention of analytic geometry should be credited to Nicole Oresme, who was born in Normandy about 1323, and who died in 1382 after a career that carried him from a mathematics professorship to a bishopric. Oresme, in one of his mathematical tracts, anticipated another aspect of analytic geometry when he represented certain laws by graphing the dependent variable against the independent one, as the latter variable was permitted to take on small increments. Advocates for Oresme as the inventor of analytic geometry see in his work such accomplishments as the first explicit introduction of the equation of a straight line and the extension of some of the notions of the subject from two-dimensional space to three-dimensional and even four-dimensional space. A century after Oresme's tract was written, it enjoyed several printings, and in this way may possibly have influenced later mathematicians.

The above views of analytic geometry seem to confuse the subject with one or more of its aspects. The real essence of the subject, however, is its transform-solve-invert character, wherein a problem in geometry is first *transformed* into a corresponding one in algebra, the algebraic problem then *solved*, and finally the algebraic solution *inverted* to obtain the geometric solution. It follows that, before analytic geometry could assume its highly practical form, it had to await the development of algebraic processes and symbolism. It would therefore seem much more correct to agree with the majority of historians, who regard the decisive contributions made in the seventeenth century by the two French mathematicians, René Descartes (1596–1650) and Pierre de Fermat (1601?-1665), as the essential origin of the subject. Certainly, not until after the impetus given to the subject by these two men do we find analytic geometry in a form with which we are familiar. The initial contributions made by these two men can be regarded as one of the greatest of the GREAT MOMENTS IN MATHEMATICS.

Descartes' claim to the invention of analytic geometry rests on one of the three appendices to his famous treatise on universal science, *Discours de la méthode pour bien conduire sa raison et chercher la vérité dans les sciences* (Discourse on the Method of Rightly Conducting Reason and Seeking Truth in the Sciences), which was published in 1637. It is the last, or third, appendix, entitled *La géométrie,* which contains Descartes' contributions to analytic geometry. This appendix starts off with an explanation of some of the principles of algebraic geometry and shows a real advance over the Greeks. To the Greeks a variable corresponded to the length of some line segment, the product of two variables to the area of some rectangle, and the product of three variables to the volume of some rectangular parallelepiped. Beyond this the Greeks could not go. To Descartes, on the other hand, x^2 did not suggest an area, but rather the fourth term in the proportion $1:x = x:x^2$, and as such is representable by an appropriate line segment, which can easily be constructed when x is known. Using a unit segment we can, in this way, represent any power of a variable or the product of any number of variables by a line length and actually construct the line length with Euclidean tools when the values of the variables are assigned. With this arithmetization of geometry, Descartes, in *La géométrie,* marks off x on a given axis and then a length y at a fixed angle to this axis and endeavors to construct points whose x and y satisfy a given relation (see Figure 67). For example, if we have the relation $y = x^2$, then for each value of x we are able to construct the corresponding y as the fourth term of the proportion given above. Descartes shows special interest in obtaining such relations for curves which are defined kinematically.

At the same time that Descartes was formulating the basis of modern analytic geometry, the subject was also occupying the attention of Pierre de Fermat. Fermat's claim to priority rests on a letter written to Roberval in September 1636, in which it is stated

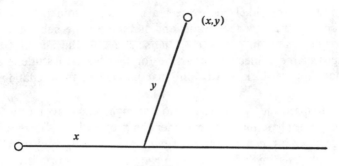

FIG. 67

that the ideas of the writer were even then seven years old. The details of the work appear in the posthumously published paper *Isogoge ad locus planos et solidos*. Here we find the equation of a general straight line and of a circle, and a discussion of hyperbolas, ellipses, and parabolas. In a work on tangents and quadratures, completed before 1637, Fermat defined many new curves analytically. Where Descartes suggested a few new curves, gener ated by mechanical motion, Fermat proposed many new ones, defined by algebraic equations. The curves $x^m y^n = a$, $y^n = ax^m$, and $r^n = a\theta$ are still known as *hyperbolas, parabolas,* and *spirals of Fermat.* Thus, where to a large extent Descartes began with a locus and then found its equation, Fermat started with the equation and then studied the locus. These are the two inverse aspects of the fundamental principle of analytic geometry.

Too frequently, introductory courses in analytic geometry are much too slim to furnish the student a true appreciation of the subject; this is particularly so in those courses where analytic geometry is reduced to just the barest minimum needed for a first study of the calculus. The result is that many college students encounter little more in analytic geometry than the simple plotting of some points and curves and a way of recognizing the shapes of conic sections from equations given in more or less standard forms. In these abbreviated treatments of analytic geometry, the student never gets to feel the remarkable power of the subject, never realizes its surprising flexibility, diversity, and applicability, and, indeed, very likely never sees a nontrivial geometry problem attacked by the new technique and never sees the subject in its role of discoverer in geometry. One sensitive to the beauties and strength of the subject must grieve for the deprived student and wonder over the lack of wisdom that has led to this pedagogical plight.

To compare the synthetic method of high school geometry with the newer analytic method, consider the proposition: *The medians of a triangle are concurrent in a point that trisects each median.* The student is invited to try to recall how this proposition was proved in his high school course or, if unable to recall the proof, to try to supply a high school demonstration of his own. There is a good chance the student will fail either way. The reason for this is that a high school demonstration of the proposition requires the drawing of some preliminary auxiliary lines, and it is not at all easy to remember or to guess just what lines to draw.

The demonstration found in most high school texts of plane geometry runs as follows (see Figure 68). Let the medians BE and CF of triangle ABC intersect in G, and let M and N denote the midpoints of BG and CG, respectively. Draw FE, MN, FM, EN. Then FE is parallel to BC and equal to one-half of BC (the line segment joining the midpoints of two sides of a triangle is parallel to the third side and is equal to one-half the third side). Similarly, MN is parallel to BC and is equal to one-half of BC. Therefore FE is both parallel and equal to MN, whence $FENM$ is a parallelogram. It follows that $MG = GE$ and $NG = GF$. That is, the two medians BE and CF intersect in a point G which is two-thirds the way from either vertex to the midpoint of the opposite side. Since this is true of any pair of medians of triangle ABC, it finally follows that all three medians are concurrent in a point that trisects each median. This common point of the three medians of the triangle is called the *centroid* of the triangle.

The preceding demonstration, it must be confessed, has certain pleasing aesthetic qualities, but one sees why it is neither easy to recall nor easy to invent on one's own—one simply does not remember or does not know what to do first (namely, draw the auxiliary lines FE, MN, FM, EN). This is the major trouble with

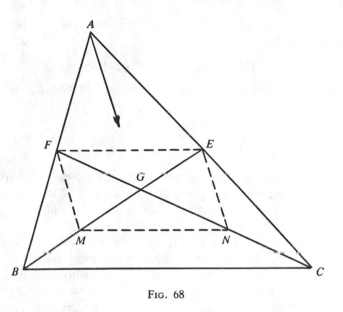

FIG. 68

the synthetic method—one does not know how to begin, for no step-by-step order of approach is dictated to the solver. The synthetic method requires an innate aptitude or a skill that comes only after much practice and experience.

Let us now re-establish the proposition about the medians of a triangle, this time by the method of analytic geometry. Place the triangle ABC anywhere on a rectangular Cartesian frame of reference (see Figure 69). Then the vertices A, B, C have coordinates (a_1, a_2), (b_1, b_2), (c_1, c_2), say, with respect to the frame. Let us find the coordinates (g_1, g_2) of the second trisection point G of median AD. All we need for this is the so-called *ratio formula* of analytic geometry, which says that the coordinates (p_1, p_2) of the point P dividing a line segment MN so that $MP/PN = r/s$ are given by

$$p_1 = (sm_1 + rn_1)/(s + r), \qquad p_2 = (sm_2 + rn_2)/(s + r),$$

where the coordinates of M and N are (m_1, m_2) and (n_1, n_2), respectively. Using the ratio formula, we find the coordinates (d_1, d_2) of D to be

$$d_1 = (b_1 + c_1)/2, \qquad d_2 = (b_2 + c_2)/2.$$

Using the ratio formula again, we find the coordinates (g_1, g_2) of G to be

$$g_1 = (a_1 + 2d_1)/3 = (a_1 + b_1 + c_1)/3,$$
$$g_2 = (a_2 + 2d_2)/3 = (a_2 + b_2 + c_2)/3.$$

In exactly the same way, or simply by symmetry, we see that G is also the second trisection point of the medians BE and CF. It follows that the three medians are concurrent in G, which trisects each median, and the proposition is established.

Examining the preceding proof, we note that we did not encounter the prime difficulty encountered when using the synthetic method. That is, there was no question as to what to do first, and then what to do next, and so on. The first thing was to place the figure on a frame of reference and assign coordinates to the given

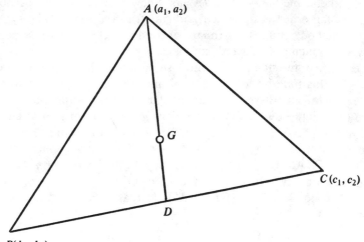

FIG. 69

points. Then, since we wanted to prove that the three medians have a common second trisection point, we set about finding the coordinates of these second trisection points. This is rather characteristic of the analytic method; we generally know pretty much just how to proceed. Of course, in order to carry out the procedure, one must have an assortment of useful analytic-geometry formulas at one's fingertips (in our example we needed the ratio formula). It is for this reason that an introductory course in analytic geometry commences with the establishment of a number of such useful formulas.

The three-dimensional analogue of the proposition just established is known as *Commandino's theorem.** We first define the lines joining the vertices of a tetrahedron to the centroids of the opposite faces to be the *medians* of the tetrahedron. Commandino's theorem then states that: *The four medians of a tetrahedron are concurrent in a point that quadrisects each median.* The student may care to try to supply a synthetic demonstration of this theorem; he certainly will notice that the analytic demonstration of the two-dimensional theorem immediately extends to the three-dimensional situation. Here is another advantage of the analytic method: an analytical proof is often easily extended to cover a higher dimensional or a more general situation.

Summarizing, we may say that an important advantage of the analytic method over the synthetic method is that with the former method we usually have a definite step-by-step procedure to follow, whereas with the latter method we must resort to experience and to "trial and error." Does this mean that the analytic method is more or less routine and that there is little room for genius on the part of the solver? No indeed, for though one may know *what* to do at a particular step of an analytical proof, it is another matter to *do* it. The algebra may be too complex to carry out; indeed, one's algebraic ability may simply not be strong enough to sustain the attack. For example, in attempting to prove some proposition by the analytic method, one may reach a point where

---

*It was given by Federigo Commandino (1509–1575) in his work *De centro gravitatis solidorum* of 1565. The proposition may have been known to the ancient Greeks.

all one has to do is to solve a certain given octic equation; but such a solution could be very difficult, if not actually impossible. Here, then, is the pitfall of analytic geometry: we often know what to do, but lack the technical ability to do it. In this case the solver must recast his approach in some clever way that will avoid the forbidding algebra, and it is here that skill and genius are required when employing the analytic method.

Experience has shown, however, that of the two methods, the synthetic and the analytic, the latter is generally the broader, the more powerful, and the easier to apply, but a good geometer certainly does not hesitate to use whichever method seems best to fit the investigation at hand.

In his *Eudemian Summary,* Proclus (410–485) tells us that Ptolemy Soter, the first King of Egypt and founder of the Alexandrian Museum, patronized the school by studying geometry there under Euclid. He found the subject difficult and one day asked his teacher, "Isn't there an easier way to do this?" Euclid replied, "Oh King, in the real world there are two kinds of roads, roads for the common people to travel upon and roads reserved for the King to travel upon. In geometry there is no royal road." Since so many students are considerably more able as algebraists than as geometers, perhaps analytic geometry is the "royal road in geometry" that Euclid thought did not exist.

There are a couple of legends describing the initial hint that led Descartes to the contemplation of analytic geometry. According to one story, it came to him in a dream. On St. Martin's Eve, November 10, 1619, while the army with which he was soldiering was lying inactive in its winter quarters on the banks of the Danube, Descartes experienced three singularly vivid and coherent dreams that, he claimed, changed the whole course of his life. The dreams, he said, clarified his purpose in life and determined his future endeavors by revealing to him "a marvelous science" and "a wonderful discovery." Descartes never explicitly disclosed just what were the marvelous science and the wonderful discovery; but it is believed to have been analytic geometry, or the application of algebra to geometry, and then the reduction of all the sciences to geometry. It wasn't until eighteen years later that he expounded some of his ideas in his famous publication of 1637.

Another story, perhaps on a par with the story of Isaac Newton and the falling apple, says that the initial flash of analytic geometry came to Descartes one morning while lying awake in bed. Watching a fly crawl about on the ceiling near a corner of his room, it struck him that the path of the fly on the ceiling could be described if only one knew the relation connecting the fly's distances from the two adjacent walls. Even though this second story may be apocryphal, it has good pedagogical value.

Exercises

20.1. Establish the following theorems:
(a) A necessary and sufficient condition for a triangle inscribed in an ellipse to have maximum area is that the centroid of the triangle coincide with the center of the ellipse.
(b) The envelope of a chord of an ellipse that cuts off a segment of constant area is a concentric similar, and similarly oriented, ellipse.

20.2.(a) List a number of formulas useful when working with a rectangular Cartesian coordinate system.
(b) Which of the formulas above continue to hold in an oblique Cartesian coordinate system?

20.3.(a) Give an analytic proof of Commandino's theorem.
(b) Give a synthetic proof of Commandino's theorem.

20.4.(a) The line segments joining the midpoints of pairs of opposite edges of a tetrahedron are called the *bimedians* of the tetrahedron. Prove, analytically, that the bimedians of a tetrahedron are all bisected by the centroid of the tetrahedron.
(b) Show that the sum of the squares of two pairs of opposite edges of a tetrahedron is equal to the sum of the squares of the remaining two opposite edges increased by four times the square of the bimedian relative to these last edges.
(c) Show that the sum of the squares of the edges of a tetrahedron is equal to four times the sum of the squares of its bimedians.

20.5. Find the locus of a point P the square of whose distance from the hypotenuse of a given right triangle is equal to the product of its distances from the two legs.

20.6. Prove, analytically, that the altitudes of a triangle are concurrent.

20.7.(a) Prove, analytically, that the midpoints of the sides of a planar quadrilateral are the vertices of a parallelogram.

(b) Is the preceding result still true if the quadrilateral is not planar?

20.8. Pirates buried a treasure on an island in the following manner: Near the shore were two large rocks and a lone palm tree. One pirate started from one of the rocks, and walking along the line at right angles to the line joining the rock to the palm tree, paced off a distance equal to that between the rock and the palm tree. A second pirate did a similar thing with respect to the other rock and the palm tree. The treasure was then buried midway between the positions occupied by the two pirates. Years later the pirates returned to the island to dig up their treasure, but found that the palm tree was no longer there. A cabin boy, who had studied elementary analytic geometry, located the treasure for the pirates. How did he do this?

20.9. Points D, E, F are selected on the sides BC, CA, AB of a given triangle ABC such that $BD/BC = CE/CA = AF/AB = \frac{1}{3}$. Show that the area of the triangle formed by the three lines AD, BE, CF is one-seventh the area of the given triangle.

20.10. The extremities A and B of a line segment AB of constant length move along two perpendicular lines. Find the locus of the point P which divides AB in a fixed ratio.

Further Reading

BOYER, C. B., *History of Analytic Geometry*. New York: *Scripta Mathematica*, Yeshiva University, 1956.

COOLIDGE, J. L., *A History of Geometrical Methods*. New York: Oxford University Press, 1940.

MAHONEY, M. S., *The Mathematical Career of Pierre de Fermat, 1601–1665*. Princeton, N.J.: Princeton University Press, 1972.

SMITH, D. E., and MARCIA L. LATHAM, eds., *The Geometry of René Descartes*. New York: Dover, 1954.

HINTS FOR THE SOLUTION OF SOME
OF THE EXERCISES

1.3. Here we have number names that originated from gestures formerly used to express the numbers.

1.4. A man and his wife sleep on the same mattress.

1.6. (b) Use the one-to-one correspondence $n \leftrightarrow 2n$, where n is any positive integer.

1.7. The properties follow immediately from the commutativity and associativity of the union operation on sets.

1.8. Let A, B, C be any three sets. The correspondence (a, b) $\leftrightarrow (b, a)$, where $a \in A$ and $b \in B$, establishes the commutative property. The correspondence $(a,(b, c)) \leftrightarrow ((a, b),c)$, where $a \in A$, $b \in B$, $c \in C$, establishes the associative property. Now show that $A \times (B \cup C) = (A \times B) \cup (A \times C)$. Then, if B and C have no element in common and if α, β, γ denote the cardinal numbers of A, B, C, the cardinal number of $A \times (B \cup C)$ is $\alpha (\beta + \gamma)$, and the cardinal number of $(A \times B) \cup (A \times C)$ is $\alpha\beta + \alpha\gamma$.

1.9. If A has n elements, then A contains 2^n subsets, because in forming a subset of A we have two choices in regard to each of the n elements of A, namely, to include the element or not to include the element.

1.10. If A has m elements and B has n elements, then $A \cap B$ has p elements, where $0 \leq p \leq \min(m,n)$ and $A \cup B$ has q elements, where $\max(m,n) \leq q \leq m + n$.

2.1. (b) Here $s = r$, $c = 2r$.

229

2.1. (c) If r is the radius of the circle and θ is half the central angle subtended by the chord, we have $r = (4s^2 + c^2)/8s$, $\theta = \sin^{-1}(c/2r)$, $A = r^2\theta - c(r - s)/2$.

2.2. (a) Set $\pi r^2 = (8d/9)^2 = (16r/9)^2$.

2.3. If r is the radius of the circle, we assume $6r/2\pi r = 57/60 + 36/3600 = 576/600$.

2.4. (a) Let the quadrilateral be $ABCD$, where $AB = a$, $BC = b$, $CD = c$, $DA = d$. Then

$$4K = ab \sin B + bc \sin C + cd \sin D + da \sin A$$
$$\leq ab + bc + cd + da = (a + c)(b + d).$$

2.4. (b) Set $d = 0$ in the formula of 2.4 (a).

2.5. Take $\pi = 3$. Then, if r is the radius of the circle, $2r = 60/3 = 20$. The sagitta s of the chord c is 2. Now $c^2 = (2r)^2 - (2r - 2s)^2$.

2.6. We have $\pi d^2 = 4s^2$.

2.7. Start with $|\sqrt{m} - \sqrt{n}| \geq 0$.

2.8. Complete the pyramid of which the frustum is a part and express the volume of the frustum as the difference between the volumes of the completed and added pyramids.

3.2. Set up a vertical stick of length s near the pyramid. Let S_1, P_1 and S_2, P_2 be the points marking the shadows of the top of the stick and the apex of the pyramid at two different times of day. Then, if x is the sought height of the pyramid, $x = s(P_1P_2)/(S_1S_2)$.

3.3. (a) Draw a line through a vertex of the triangle parallel to the opposite side.

3.4. (g) Use limit concepts applied to 3.4 (f).

3.6. Fold the vertices onto the foot of an altitude, or onto the incenter of the triangle. There is reason to believe that Pascal employed the latter method.

3.9. The formula holds for all *convex* polyhedra and, more generally, for all polyhedra continuously deformable into a sphere.

3.10. The list cannot be continued; there is no convex polyhedron all of whose faces are hexagons. In fact, it can be shown that any convex polyhedron must have some face with less than six sides.

3.11. (b) Let D be a point on the surface of P such that the distance CD is a minimum. Show that D can be neither a vertex of P nor lie on an edge of P and that CD is perpendicular to the face F of P on which D lies.

3.12. Neither one gives E. The rectification of an ellipse involves a Legendre ellliptic integral of the second kind and is not expressible in terms of elementary functions.

3.13. See Problem E753, *The American Mathematical Monthly*, Aug.–Sept. 1947.

3.14. No. Only for so-called *orthocentric tetrahedra* are the four altitudes concurrent. An orthocentric tetrahedron is a tetrahedron each edge of which is perpendicular to its opposite edge.

4.1. See Figure 70.

4.2. See Figure 71.

4.4. (b) Note that the square on the shorter leg is not dissected.

4.5. (b) The four right triangles having legs of lengths 3 and 4, along with the small unit square, form a square whose area is 25.

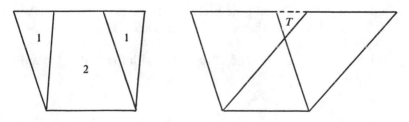

Fig. 70

It follows that the hypotenuse of a right triangle having legs 3 and 4 is 5. Since a triangle is determined by its three sides, it now follows that a 3-4-5 triangle is a right triangle.

4.7. (a) Place the tetrahedron in the first octant with the vertex of the right angle at the origin and the vertices of the base at $A = (a, 0, 0)$, $B = (0, b, 0)$, $C = (0, 0, c)$. An equation of plane ABC is then

$$bcx + cay + abz = abc$$

and the altitude h is given by

$$h = abc/(b^2c^2 + c^2a^2 + a^2b^2)^{1/2}.$$

It follows that

$$1/h^2 = (b^2c^2 + c^2a^2 + a^2b^2)/a^2b^2c^2 = 1/a^2 + 1/b^2 + 1/c^2.$$

(For a proof by synthetic geometry, see, e.g., Nathan Altshiller-Court, *Modern Pure Solid Geometry,* 2nd ed., Chelsea, New York, 1964, Theorem 285, p. 101.

4.7. (b) The area of the base ABC is given by

$$ABC = 3(\text{vol. of tetrahedron})/h = abc/2h.$$

It follows that

$$\begin{aligned}(ABC)^2 &= a^2b^2c^2/4h^2 \\ &= (a^2b^2c^2/4)(1/a^2 + 1/b^2 + 1/c^2) \\ &= (b^2c^2 + c^2a^2 + a^2b^2)/4 \\ &= (OBC)^2 + (OCA)^2 + (OAB)^2.\end{aligned}$$

4.8. Circle $AB'C$ is tangent to AB at A; circle $AC'B$ is tangent to AC at A. It follows that $(AB)^2 = (BC)(BB')$, $(AC)^2 = (BC)(CC')$. Etc.

When $\angle A = 90°$, B' and C' coincide with the foot of the altitude from A.

4.9. $\cos c = \cos a \cos b$.

4.10. If, in a triangle with sides a, b, c, we have $a^2 + b^2 = c^2$, then the triangle is a right triangle with hypotenuse c.

Let d be the hypotenuse of the right triangle with legs a and b.

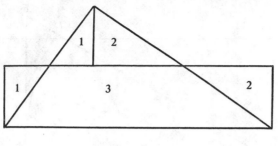

Fig. 71

Then $d^2 = a^2 + b^2 = c^2$, whence $d = c$. It follows that the given triangle is congruent to the right triangle.

5.1. (b) Let (see Figure 72) AC and BC be commensurable with respect to AP. Show that then DE and DB are also commensurable with respect to AP, and so on.

5.2. (a) If the line should pass through the point (p, q) of the coordinate lattice, we would have $\sqrt{2} = q/p$, a rational number.

5.3. Assume $\sqrt[n]{p} = a/b$, where a and b are relatively prime integers.

5.4. (a) Assume $\log_{10}2 = p/q$, where p and q are integers. Then we must have $10^p = 2^q$, which is impossible.

5.5. (a) Assume the sum is a rational number.

5.5. (b) Assume the product is a rational number.

5.6. (a) In Figure 73, isosceles triangles DAC and GDC are similar. Therefore $AD:DG = DC:GC$, whence $DB:DG = DG:GB$.

5.6. (b) $AG:AH = AG:GB = AB:AG = AB - AG:AG - AH = GB:HG = AH:HG$.

5.6. (c) Let a and b denote the length and width of the given rectangle. Then

$$\frac{a - b}{b} = \frac{a}{b} - 1 = \frac{2}{\sqrt{5} - 1} - 1 = \cdots = \frac{\sqrt{5} - 1}{2}.$$

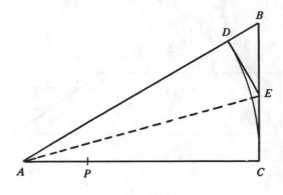

Fig. 72

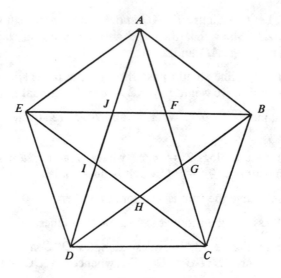

Fig. 73

5.7. (a) Let *HG* in Figure 73 be the given side. Draw a right triangle *PQR* with legs *PR* and *QR* equal to *HG* and *HG*/2, respectively. On *PQ* produced mark off *QT* = *QR*. Then *PT* = *GB* = *GC* = *HC*, and so on.

5.7. (b) Let DB in Figure 73 be the given diagonal. Draw a right triangle PQR with legs PR and QR equal to $DB/2$ and DB, respectively. On PQ mark off $PT = PR$. Then $TQ = DG = DC$, and so on.

5.7. (c) $360°/15 = 24° = 2(72° - 60°)$.

5.7. (d) Since r and s are relatively prime, there exist positive integers p and q such that $pr - qs = \pm1$. Therefore the difference between the angle subtended at the center of the s-gon by p of its sides and the angle subtended at the center of the r-gon by q of its sides is

$$p(360°/s) - q(360°/r) = (pr - qs)(360°/rs) = \pm360°/rs.$$

5.7. (e) Let p, d, h denote the lengths of sides of the regular pentagon, decagon, and hexagon inscribed in a unit circle. Then $h = 1$, $d = (\sqrt{5} - 1)/2$. In an isosceles triangle of legs 1 and base d, let t denote the altitude on one of the legs and m the projection of the base on this leg. Then

$$m^2 = d^2 - t^2, \qquad t^2 = 1 - (1 - m)^2,$$

from which we obtain $p^2 = 4t^2 = 4d^2 - d^4$. Now show that $p^2 = h^2 + d^2$.

5.8. (a) The decimal division of one integer by another leads, in the division algorithm, to a sequence of remainders. If we should reach a zero remainder, then the decimal expansion terminates. Otherwise we obtain an endless sequence of nonzero remainders each less than the divisor. Since there are only a finite number of such possible remainders, we will, sooner or later, reach a remainder obtained earlier, leading to a repeated decimal expansion.

5.8. (c) Set $x = 3.2\overline{39}$ and $y = 0.\overline{39}$. Then $10x = 32 + y$, $1000x = 3239 + y$. Eliminating y we find $990x = 3207$, or $x = 3207/990$.

5.9. (a) The expansion neither terminates nor repeats.

5.9. (b) The expansion neither terminates nor repeats.

6.2. (a) Use the fact that if a set of parallel lines cuts off equal segments on one transversal then it cuts off equal segments on any other transversal.

6.2. (b) Let the triangle be ABC and let MN, M on AB and N on AC, be parallel to side BC. Now $\triangle MNB = \triangle NMC$, since they have the common side MN and equal altitudes on this side. We now have, using the theorem established in the lecture text,

$$AM:MB = \triangle ANM:\triangle MNB = \triangle AMN:\triangle NMC = AN:NC.$$

6.4. (a) No; condition (3) does not hold.

6.4. (b) No; condition (2) does not hold.

6.4. (c) No; condition (1) does not hold.

6.4. (d) Yes.

6.4. (e) Yes.

6.5. Employ *reductio ad absurdum*.

6.6. (a) Let p_n denote the perimeter of a regular n-gon inscribed in the circle. Define the circumference c of the circle to be $\lim_{n \to \infty} p_n$.

6.6. (b) Let a_n denote the area of a regular n-gon inscribed in the circle. Define the area A of the circle to be $\lim_{n \to \infty} a_n$.

6.7. π and $\pi/4$.

6.8. (a) Let P be the given point and Q a neighboring point on the curve. Define the tangent PT to the curve at P as the limiting position, if it exists, of the secant line PQ as Q moves along the curve and approaches P as a limit.

6.8. (b) Let c be a curve on the surface S and passing through the given point P of S and let PT be the tangent, if it exists, to c at P. Under sufficient conditions of smoothness of S at P, it can be shown that all such tangents to curves c on S through P exist and lie in a plane. This plane is the tangent plane to S at P.

7.4. The janitor repaired the chairs.

7.6. By part (B) of the pattern of material axiomatics, a postulate is to be readily acceptable as true in view of the preliminary explanations given of the primitive terms of the discourse.

7.7. Use the fact that the number of combinations of four things taken two at a time is 6.

7.8. Examine all possible cases.

7.9. Let the first player place his first cigar so that its center is directly over the center of the table, and let him thenceforth place his cigars in the positions centrally symmetrical to those occupied by the second player's cigars.

7.10. In chess, white has the opening move. After A moves, B copies A's move in the game against C on the other board. When C answers B's move, B copies it as his answer to A's move in the game on the first board.

8.1. Choose 2 and 3.

8.4. (a) 73.

8.4. (b) 29.

8.4. (c) Suppose $a > b$. Then the algorithm may be summarized as follows:

$$
\begin{aligned}
a &= q_1 b + r_1 & & 0 < r_1 < b \\
b &= q_2 r_1 + r_2 & & 0 < r_2 < r_1 \\
r_1 &= q_3 r_2 + r_3 & & 0 < r_3 < r_2 \\
&\cdots\cdots & & \cdots\cdots \\
r_{n-2} &= q_n r_{n-1} + r_n & & 0 < r_n < r_{n-1} \\
r_{n-1} &= q_{n+1} r_n
\end{aligned}
$$

Now, from the last step, r_n divides r_{n-1}. From next to the last step r_n divides r_{n-2}, since it divides both terms on the right. Similarly, r_n divides r_{n-3}. Successively, r_n divides each r_i, and finally a and b.

On the other hand, from the first step, any common divisor c of a and b divides r_1. From the second step, c divides r_2. Successively, c divides each r_i. Thus c divides r_n.

8.4. (d) From the next to last step in the algorithm we can express r_n linearly in terms of r_{n-1} and r_{n-2}. From the preceding step we can then express r_n linearly in terms of r_{n-2} and r_{n-3}. Continuing this way we finally obtain r_n linearly in terms of a and b

8.4. (e) $-7(7592) + 9(5913) = 73$.

8.4. (f) a and b are relatively prime if and only if g.c.d. $= h = 1$.

8.5. (a) If p does not divide u, then integers P and Q exist such that $Pp + Qu = 1$, or $Ppv + Quv = v$.

8.5. (b) Suppose there are two prime factorizations of the integer n. If p is one of the prime factors in the first factorization it must, by part (a), divide one of the factors in the second factorization, that is, it must equal one of those factors.

8.5. (c) Note that $273 = (13)(21)$. Find [see Exercise 8.4 (f)] integers p and q such that $13p + 21q = 1$. Dividing by 273 we then have $p/21 + q/13 = 1/273$. Similarly find integers r and s such that $1/21 = r/3 + s/7$.

8.6. (c) For each b_i in part (b) may take on $a_i + 1$ values.

8.6. (d) a is a perfect square if and only if each a_i is even; each a_i is even if and only if the product in part (c) is odd.

8.6. (f) Since b divides ac, we have $b_i \leq a_i + c_i$. Also, since a and b are relatively prime, we have $a_i = 0$ or $b_i = 0$. In either case $b_i \leq c_i$.

8.6. (g) Since a divides c, $a_i \leq c_i$. Since b divides c, $b_i \leq c_i$. Since a and b are relatively prime, $a_i = 0$ or $b_i = 0$. It follows that $a_i + b_i \leq c_i$.

8.6. (h) Suppose $\sqrt{2} = a/b$, where a and b are positive integers. Then, since $a^2 = 2b^2$, we have $(2a_1, 2a_2, \ldots) = (1 + 2b_1, 2b_2, \ldots)$, whence $2a_1 = 1 + 2b_1$, which is impossible.

8.7. To Euclid a circle was a circular disc.

8.8. (b) Let A be the given point and BC the given line segment. Construct, by I 1, an equilateral triangle ABD. Draw circle $B(C)$ and let DB produced cut this circle in G. Now draw circle $D(G)$ to cut DA produced in L. Then AL is the sought segment.

8.9. (a) See Figure 74.

8.9. (b) See Figure 75.

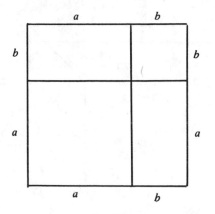

Fig. 74

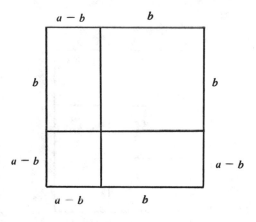

Fig. 75

8.9. (c) See Figure 76.

8.9. (d) See Figure 77.

8.10. (a) We have $(x - r)(x - s) = x^2 - px + q^2$.

8.10. (h) Denote the parts by x and $a - x$. Then $x^2 - (a - x)^2 = x(a - x)$, or $x^2 + ax - a^2 = 0$.

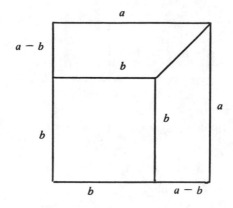

FIG. 76

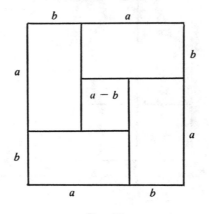

FIG. 77

9.1. (a) w_1 pounds of pure gold will lose $w_1 f_1/w$ pounds when weighed in water; w_2 pounds of pure silver will lose $w_2 f_2/w$ pounds when weighed in water. It follows that

$$w_1 f_1/w + w_2 f_2/w = f.$$

9.1. (b) $f:f_1:f_2 = v:v_1:v_2$.

9.2. $(\pi r^2)(2r) = (3/2)V$, $(2\pi r)(2r) = (3/2)A$.

9.3. Consult a high school textbook on solid geometry.

9.4. If a is the chord of the generating arc, then $a^2 = (2r)(h)$.

9.5. (a) The volume of the segment is equal to the volume of a spherical sector minus the volume of a cone. Also, $a^2 = h(2r - h)$.

9.5. (b) The segment is the difference of two segments, each of one base, and having, say, altitudes u and v. Then

$$V = \pi r(u^2 - v^2) - \frac{\pi(u^3 - v^3)}{3}$$

$$= \pi h \left[(ru + rv) - \frac{u^2 + uv + v^2}{3} \right].$$

But $u^2 + uv + v^2 = h^2 + 3uv$ and also $(2r - u)u = a^2$ and $(2r - v)v = b^2$. Therefore

$$V = \pi h \left(\frac{a^2 + b^2}{2} + \frac{u^2 + v^2}{2} - \frac{h^2}{3} - uv \right)$$

$$= \pi h \left(\frac{a^2 + b^2}{2} + \frac{h^2}{2} + uv - \frac{h^2}{3} - uv \right), \text{ etc.}$$

9.6. (a) We have $(OM)(AO) = (OP)(AC)$. Summing we then find

$$\text{(area of segment)}(HK) - (\triangle AFC)(KC/3).$$

9.6. (b) $\triangle AVC = (\triangle AWC)/2 = (\triangle AFC)/4$.

9.7. Draw CO and use the fact that an exterior angle of a triangle is equal to the sum of the two remote interior angles.

9.8. In Figure 78, $OP = $ arc AB. It follows that if we take OP perpendicular to OA, then OP will have a length equal to one fourth the circumference of the circle. Since the area K of the circle is half the product of its radius and its circumference, we have

$$K = (a/2)(4OP) = (2a)(OP),$$

and the side of the required square is thus the mean proportional between $2a$ and OP, or between the diameter of the circle and the length of that radius vector of the spiral which is perpendicular to OA.

9.9. Let OB cut the spiral in P and trisect the segment OP by points P_1 and P_2. If circles $O(P_1)$ and $O(P_2)$ cut the spiral in T_1 and T_2, then OT_1 and OT_2 trisect the angle AOB.

9.10. (a) Produce CB to E so that $BE = BA$. Prove triangles MBA and MBE are congruent.

10.3. (a) Take a side of the triangle as one unit and apply Ptolemy's theorem to quadrilateral $PACB$.

10.3. (b) Take a side of the square as one unit and apply Ptolemy's theorem to quadrilaterals $PBCD$ and $PCDA$.

10.3. (c) Take a side of the pentagon as one unit and apply Ptolemy's theorem to the quadrilaterals $PCDE$, $PCDA$, $PBCD$.

10.3. (d) Take a side of the hexagon as one unit and apply Ptolemy's theorem to quadrilaterals $PBCD$, $PEFA$, $PBCF$, $PCFA$.

10.4. (a) Suppose (see Figure 79) D, E, F are collinear on a line m. Drop perpendiculars p, q, r on m from A, B, C. Then, disregarding signs,

$$BD/DC = q/r, \quad CE/EA = r/p, \quad AF/FB = p/q.$$

It follows, now letting all segments be directed segments, that

$$(BD/DC)(CE/EA)(AF/FB) = \pm 1.$$

Since, however, m must cut one or all three sides externally, we see that we can have only the $-$ sign.
Conversely, suppose

$$(BD/DC)(CE/EA)(AF/FB) = -1$$

and let EF cut BC in D'. Then D' is a menelaus point and, by the above,

$$(BD'/D'C)(CE/EA)(AF/FB) = -1.$$

It follows that $BD/DC = BD'/D'C$, or that $D \equiv D'$. That is, D, E, F are collinear.

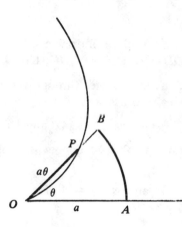

Fɪɢ. 78

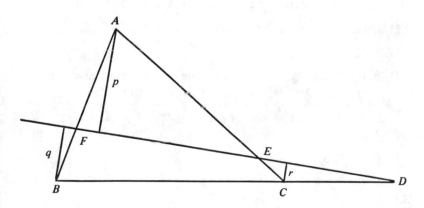

Fɪɢ. 79

10.4. (b) Let h denote the length of the perpendicular from O to line BC. The reader may then check that for all possible figures

$$\frac{BD}{DC} = \frac{h\ BD}{h\ DC} = \frac{2\triangle OBD}{2\triangle ODC} =$$

$$\frac{(OB)(OD)\sin BOD}{(OD)(OC)\sin DOC} = \frac{OB\ \sin BOD}{OC\ \sin DOC}.$$

10.4. (c) Use 10.4 (b).

10.4. (d) In the figure for 10.4 (c), draw a sphere with center O to cut OA, OB, OC, OD, OE, OF in A', B', C', D', E', F'.

10.5. (a) Draw the circumdiameter through the vertex through which the altitude passes, and use similar triangles.

10.5. (b) Apply 10.5 (a) to triangles DAB and DCB.

10.5. (c) Use the result of 10.5 (b) along with Ptolemy's relation $mn = ac + bd$.

10.5. (d) Here $\theta = 0°$ and $\cos \theta = 1$. Now use 10.5 (b) and 10.5 (c).

10.5. (e) Use 10.5 (a).

11.1. (b) For $n = 2$ we obtain the Pythagorean pair 220 and 284; for $n = 4$ we obtain the Fermat pair 17296 and 18416.

11.2. (a) Show that $2^{ab} - 1$ contains the factor $2^a - 1$.

11.2. (b) 8128.

11.2. (c) If a_1, a_2, ..., a_n represent the divisors of N, then N/a_1, N/a_2, ..., N/a_n also represent the divisors of N.

11.3. (a) The sum of the proper divisors of p^n is $(p^n - 1)/(p - 1)$.

11.4. (a) 1, 6, 15, 28.

11.4. (b) $T_n = 1 + 2 + \cdots + n$, $P_n = 1 + 4 + 7 + \cdots + (3n - 2)$.

11.4. (c) Use 11.4 (b).

11.4. (d) An oblong number is of the form $a(a + 1)$.

11.4. (f) See Figure 80.

11.4. (g) $2^{n-1} (2^n - 1) = 2^n (2^n - 1)/2$. Now use 11.4 (b).

11.4. (h) $a = (m - 2)/2$, $b = (4 - m)/2$.

11.4. (i) $a = 5/2$, $b = -3/2$.

11.5. (b) If p is composite, then $p = ab$, where $a \le b$ and, consequently, $a^2 \le p$.

11.5 (c) For $n = 10^9$ we have $(A_n \log_e n)/n = 1.053 \ldots$.

11.5. (d) Consider $(n + 1)! + 2$, $(n + 1)! + 3$, $\ldots$, $(n + 1)!$ $+ (n + 1)$.

11.6. (a) $(2m)^2 + (m^2 - 1)^2 = (m^2 + 1)^2$.

11.6. (b) If there were an isosceles right triangle with integral sides, then $\sqrt{2}$ would be rational.

11.6. (c) If there are positive integers a, b, c, $(a \ne 1)$ such that $a^2 + b^2 = c^2$ and $b^2 = ac$, then a, b, c cannot be relatively prime. But if there is a Pythagorean triple in which one integer is a mean proportional between the other two, there must be a primitive Pythagorean triple of this sort.

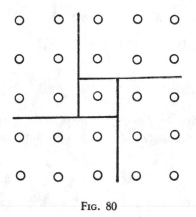

FIG. 80

11.6. (e) Show that if $a^2 + (a + 1)^2 = c^2$, then

$$(3a + 2c + 1)^2 + (3a + 2c + 2)^2 = (4a + 3c + 2)^2.$$

11.6. (f) Use 11.6 (e).

11.6. (g) If n is odd and > 2, $(n, (n^2 - 1)/2, (n^2 + 1)/2)$ is a Pythagorean triple. If n is even and > 2, $(n, n^2/4 - 1, n^2/4 + 1)$ is a Pythagorean triple.

11.6. (h) Since $a^2 = (c - b)(c + b)$ it follows that $b + c$ is a factor of a^2. Therefore $b < a^2$ and $c < a^2$, and the number of combinations of such natural numbers b and c is finite.

11.7. (a) 7, 4, 11, 9.

11.7. (b) Set $CD = 3x$, $AC = 4x$, $AD = 5x$, $CB = 3y$. Then, since $AB/DB = AC/CD$, we find $AB = 4(y - x)$. By the Pythagorean theorem we are led to $7y = 32x$. We finally get $AB = 100$, $AD = 35$, $AC = 28$, $BD = 75$, $DC = 21$.

11.7. (d) 1806.

11.8. (b) It is easily shown that $x = x_1 + mb$ and $y = y_1 - ma$ constitute a solution. Conversely, assume x and y form a solution. Then $a(x - x_1) = b(y_1 - y)$, or $x - x_1 = mb$ and $y_1 - y = ma$.

11.8. (c) Dividing by 7 we find

$$x + 2y + (2/7)y = 29 + 6/7.$$

Therefore there exists an integer z such that

$$(2/7)y + z = 6/7,$$

or

$$2y + 7z = 6.$$

This can be solved by inspection to give $z_1 = 0$, $y_1 = 3$. Then $x_1 = 23$. The general solution of the original equation is then, by 11.8 (b),

$$x = 23 + 16m, \qquad y = 3 - 7m.$$

Since, by requirement, $x > 0$, $y > 0$, we must have $m \geq -1$ and

$m \leq 0$. The only permissible values of m are 0 and -1. We thus get two solutions,

$$x = 23, \quad y = 3 \quad \text{and} \quad x = 7, y = 10.$$

Or find, as in Exercise 8.4 (f), a and b such that $7a + 16b = 1$. Then we may take $x_1 = 209a$ and $y_1 = 209b$.

11.8. (d) There are four solutions: $x = 124, y = 4$; $x = 87$, $y = 27$; $x = 50, y = 50$; $x = 13, y = 73$.

11.8. (e) Let x represent the number of dimes and y the number of quarters. Then we must have $10x + 25y = 500$.

11.9. Let x denote the number of fruits in a pile and y the number of fruits each traveler receives. Then we have $63x + 7 = 23y$. The smallest permissible value for x is 5.

11.10. (a) Let $n = ab$. Then, if $x^n + y^n = z^n$, we have $(x^a)^b + (y^a)^b = (z^a)^b$.

11.10. (b) Suppose the point $(a/b, c/d)$, where a, b, c, d are integers, is on the curve. Then $(ad)^n + (bc)^n = (bd)^n$.

11.10. (c) Consider the right triangle whose sides are given by

$$a = 2mn, \ b = m^2 - n^2, c = m^2 + n^2.$$

The area of this triangle is

$$A = (ab)/2 = mn(m^2 - n^2).$$

Taking $m = x^2$ and $n = y^2$, and setting $x^4 - y^4 = z^2$, we find

$$A = x^2 y^2 (x^4 - y^4) = x^2 y^2 z^2.$$

Therefore, if $x^4 - y^4 = z^2$ has a solution in positive integers x, y, z, there exists an integral-sided right triangle whose area is a square number.

Finally, if $x^4 + y^4 = z^4$, then $z^4 - y^4 = (x^2)^2$.

12.1. 1. 120 apples; 2. 60 years old; 3. 2/5 of a day; 4. 144/37 hours.

12.2. (b) 30.5 minae of gold, 9.5 minae of copper, 14.5 minae of tin, and 5.5 minae of iron.

12.3. 84 years old.

12.4. (a) 27.

12.4. (b) $5{,}780 = \epsilon'\psi\pi$.
$72{,}803 = \varsigma M\beta'\,\omega\gamma$.
$450{,}082 = \mu M\epsilon M\pi\beta$.
$3{,}257{,}888 = \tau M\kappa M\epsilon M\varsigma'\,\omega\pi\eta$.

12.5. (a) $\Delta^{T}\Delta\beta\Delta^{T}\iota\beta \overset{o}{M} \lambda\gamma \wedge K^{T}\kappa\alpha\varsigma\,\varsigma$.

12.5. (b) $y\bar{a}\ k\bar{a}$ 3 $bha\ y\bar{a}$ 2 $bha\ k\bar{a}$ 2 $bha\ ka$ 13 $r\bar{u}$ 8̇.

12.6. (a) $R\,q \lfloor R\,c \lfloor R\,q$ 68 p 2 $\rfloor m\,R\,c \lfloor R\,q$ 68 m 2 $\rfloor\rfloor$.

12.6. (b) $\sqrt[3]{4 + \sqrt{-11}} \;\; + \sqrt[3]{4 - \sqrt{-11}}$.

12.7. A cub $- B$ 3 in A quad $+ C$ plano 4 in A aequatur D solido 2.

13.1. MMMMCCCCLXVII, DCCCLXXXIX.

13.3. (b) $(3660)_7$.

13.3. (c) $(1{,}254{,}626)_7$.

13.4. (b) $(43{,}239)_{12}$.

13.4. (c) $(179{,}578{,}432)_{12}$.

13.5. (a) $(653)_8$.

13.5. (b) 9, 8, 7.

13.5. (c) No. Yes. Yes. No.

13.5. (d) We must have $79 = b^2 + 4b + 2$, whence $b = 7$.

13.5. (e) We must have $72 = 2b^3 + 2b^2$, whence $b = 3$.

13.6. (a) Denoting the digits by a, b, c we have

$$49\,a + 7b + c = 81c + 9b + a,$$

where a, b, c are less than 7.

13.6. (b) We must have $2b^2 + 1 = t^2$, t and b positive integers, $b > 3$.

13.6. (c) $b^2 + 2b + 1 = (b + 1)^2$.

13.6. (d) $(2b^2 + 2b + 1)(2b^2 - 2b + 1) = 4b^4 + 1$.

13.7. (a) Express w in the binary scale.

13.8. (a) Let t be the tens digit and u the units digit. Following instructions we have

$$2(5t + 7) + u = (10t + u) + 14$$

as the announced final result. The trick is now obvious.

13.8. (b) Let h, t, u denote the hundreds, the tens, and the units digits. Following instructions we have

$$5\{2[5(2h + 3) + 7 + t] + 3\} + u = 100h + 10t + u + 235$$

as the announced final result. The trick is now obvious.

13.8. (c) $h = 9 - u$, $t = 9$.

14.3. Take any point D' on BA. Then take E'' on CA such that $CE'' = BD'$. Let circle $D'(B)$ cut the parallel to BC through E'' in E'. Draw a line through E' parallel to AC to cut BA in A' and BC in C'. We now have a figure similar to the desired figure and having B as center of similitude.

14.4. (a) Here $p = 1$, $q = 870$.

14.5. (b) Set $x = 2y$.

14.5. (c) Eliminate x and y, obtaining a cubic equation in z.

14.5. (d) Take the cubic in x with unit leading coefficient and subject it to a linear transformation of the type $x = y + m$. Determine m so that the resulting cubic in y lacks the linear term. Etc.

14.6. Find z such that $b/a = a/z$, then m such that $c/z = a/m$.

14.7. (a) The positive roots are 2 and 4.

14.7. (b) The negative root is -1.

14.8. (a) The real roots are given by the abscissas of the points of intersection of the line $ay + bx + c = 0$ with the cubic curve $y = x^3$.

14.8. (b) $x = 1.7+$.

14.8. (c) $x = -3.5, 1, 2.5$.

14.8. (e) $x = -6, -2, -1$.

15.1. (a) We indicate a proof of the theorem for the four-digit number N having a, b, c, d for its thousands, hundreds, tens, and units digits; the proof is easily generalized. Now

$$N = 1000a + 100b + 10c + d.$$

Let $S = a + b + c + d$. Then

$$N = 999a + 99b + 9c + S = 9(111a + 11b + c) + S,$$

and so on.

15.1. (b) Let M and N be any two numbers with excesses e and f. Then there exist integers m and n such that

$$M = 9m + e \qquad N = 9n + f.$$

Now

$$M + N = 9(m + n) + (e + f)$$

and

$$MN = 9(9mn + ne + mf) + ef,$$

and so on.

15.2. (a) Let M be the given number and N that obtained by some permutation of the digits of M. Then, since M and N consist of the same digits, they have [by 15.1 (a)] the same excess e. Thus we have

$$M = 9m + e, \qquad N = 9n + e$$

and $M - N = 9(m - n)$.

15.2. (b) By 15.2 (a), the final product must be divisible by 9, whence, by 15.1 (a), the excess for the sum of the digits in the product must be 0.

15.3. Use similar triangles.

15.4. (a) $x = 2.3696$.

15.4. (b) $x = 4.4934$.

15.6. (a) Use mathematical induction. Assume the relation true for $n = k$. Then

$$
\begin{aligned}
u_{k+2}u_k &= (u_{k+1} + u_k)u_k \\
&= u_{k+1}u_k + u_k^2 \\
&= u_{k+1}u_k + u_{k+1}u_{k-1} - (-1)^k \\
&= u_{k+1}(u_k + u_{k-1}) + (-1)^{k+1} \\
&= u_{k+1}^2 + (-1)^{k+1},
\end{aligned}
$$

and so on. Or use the expression for u_n given in 15.6 (b).

15.6. (b) Set $v_n = [(1 + \sqrt{5})^n - (1 - \sqrt{5})^n]/2^n \sqrt{5}$. Show that $v_n + v_{n+1} = v_{n+2}$ and that $v_1 = v_2 = 1$. Then $v_n = u_n$.

15.6. (c) Replace u_n and u_{n+1} by the expressions for them given by 15.6 (b); divide numerator and denominator by $(1 + \sqrt{5})^n$; find the limit as $n \to \infty$.

15.6. (d) Use the relation given in 15.6 (a).

15.7. (a) A has 121/17 denarii and B has 167/17 denarii.

15.7. (b) 33 days. This may be solved as a problem in variation.

15.8. Let x represent the value of the estate and y the amount received by each son. The first son receives $1 + (x - 1)/7$, and the second receives

$$
2 + \frac{x - \left(1 + \dfrac{x - 1}{7}\right) - 2}{7}.
$$

Equating these we find $x = 36$, $y = 6$, and the number of sons was $36/6 = 6$.

15.10. 382 apples.

16.1. (b) $H = (3ac - b^2)/9a^2$, $G = (2b^3 - 9abc + 27a^2d)/27a^3$.

16.3. $x = 4$. The other two roots are imaginary.

16.4. One finds (see the lecture text) $a = 6/b$, $c = b^3/6$, whence

$$6/b + b + b^3/6 = 10,$$

etc.

16.5. $y^3 + 15y^2 + 36y = 450$.

16.6. Here $\sqrt{(n/2)^2 + (m/3)^3} = \sqrt{-2700} = 30\sqrt{-3}$.

16.7. We find $16x^2 = x^4 + 2x^3 + 3x^2 + 2x + 1 = (x^2 + x + 1)^2$, whence

$$x^2 + x + 1 = \pm 4x.$$

It follows that $x = (3 \pm \sqrt{5})/2$, $(-5 \pm \sqrt{21})/2$.

16.8. We find $y = 3$ (or -7), whence $x = 4$. The other two roots are $-2 \pm 5\sqrt{-3}$.

16.9. $y^6 - 6y^4 - 144y^2 = 2736$.

16.10. (a) We find

$$m + h - k^2 = b, \qquad k(m - h) = c, \qquad hm = d.$$

It follows, from the first two of these relations, that

$$2m = (k^3 + bk + c)/k, \qquad 2h = (k^3 + bk - c)/k.$$

From the third relation we then obtain

$$(k^3 + bk + c)(k^3 + bk - c) = 4dk^2,$$

or

$$k^6 + 2bk^4 + (b^2 - 4d)k^2 - c^2 = 0,$$

a cubic in k^2.

16.10. (b) For the quartic equation

$$x^4 - 2x^2 + 8x - 3 = 0$$

the associated cubic in k^2 is

$$k^6 - 4k^4 + 16k^2 - 64 = 0,$$

one of whose roots is given as $k^2 = 4$. It follows that $k = 2$, from which we find

$$m = (k^3 + bk + c)/2k = 3, \qquad h = (k^3 + bk - c)/2k = -1,$$

and

$$x^4 - 2x^2 + 8x - 3 = (x^2 + 2x - 1)(x^2 - 2x + 3).$$

It follows that the four roots of the original quartic equation are

$$-1 \pm \sqrt{2} \quad \text{and} \quad 1 \pm \sqrt{-2}.$$

17.1. (a) Set $\log_b mn = p$, $\log_b m = q$, $\log_b n = r$. Then $b^p = mn$, $b^q = m$, $b^r = n$. It follows that $b^p = mn = b^q b^r = b^{q+r}$, whence $p = q + r$.

17.2. (a) Set $y = \log_a N$, $z = \log_b N$, $w = \log_b a$. Then $a^y = N$, $b^z = N$, $b^w = a$. Therefore $b = a^{1/w}$, or $b^z = a^{z/w} = a^y$. Thus $y = z/w$.

17.2. (b) Set $y = \log_b N$ and $z = \log_N b$. Then $b^y = N$, $N^z = b$, whence $n = b^{1/z} = b^y$. Thus $y = 1/z$.

17.2. (c) Set $y = \log_N b$ and $z = \log_{1/N}(1/b)$. Then $N^y = b$, $(1/N)^z = 1/b$, whence $N = b^{1/z} = b^{1/y}$. Thus $y = z$.

17.3. (a) $\log 4.26 = 1/2 + 1/8 + 1/256 + \cdots = 0.6294 \ldots$.

17.4. (b) $\cos c = \cos a \cos b$.

17.5. (a) $A = 122° \, 39'$, $C = 83° \, 5'$, $b = 109° \, 22'$.

17.5. (b) $A = 105° \, 36'$, $B = 44° \, 0'$, $c = 78° \, 46'$.

17.6. (a) Popsicle sticks or tongue depressors make excellent rods.

17.6. (b) To divide 589475 by 1615, for example, place rods headed by 1, 6, 1, 5 side by side, as shown in Figure 58, and now use these to find the successive partial quotients of the division algorithm.

17.7. (b) Refer to the laws of logarithms expressed in Exercises 17.1 (a) and 17.1 (b).

17.8. Set the A scale directly above the D scale and refer to the law of logarithms expressed in Exercise 17.1 (d) with $s = 2$.

17.9. Construct a logarithmic scale exactly one-third as long as the D scale and designate by K three of these short scales placed in tandem. Now set the K scale directly above the D scale and refer to the law of logarithms expressed in Exercise 17.1 (d) with $s = 3$.

18.1. (a) Acceleration is the increase in velocity during a unit interval of time.

18.2. (a) Open the compasses so that the given segment, or some simple fractional part of it, AA' stretches between the 100 marks on the two scales of the compasses (see Figure 81). Then the distance between the two 20 marks is one-fifth of AA'.

18.2. (b) Open the compasses so that AA'/OA in Figure 81 is the desired ratio of scale. Then BB' is the new length to be associated with the old length OB.

18.3. (a) Connect a on one arm to b on the other. Through c on the first arm draw a parallel to the line just drawn to cut the other arm in the sought fourth proportional.

18.3. (b) Open the compasses so that the distance between the 106 marks is equal to 150. Then the distance between the 100 marks represents the amount of the investment one year ago. Perform the operation five times to find the required amount.

18.4. The small circle slides as well as rolls.

18.5. Use the one-to-one correspondence $n \leftrightarrow n^2$.

18.6. (a) At perihelion.

18.6. (c) 1000 years.

18.6. (d) 25 A.U.

18.7. (a) The periods are the same.

18.7. (b) 1 hour and 24 minutes.

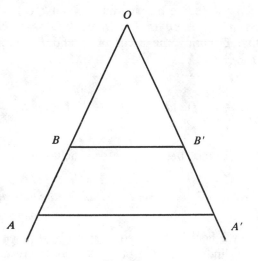

FIG. 81

18.8. (a) The interior angle at each vertex of a regular n-gon is

$$(n - 2)180°/n.$$

If there are p regular n-gons clustered about each vertex, then

$$p(n - 2)180°/n = 360°,$$

or $p = 2 + 4/(n - 2)$.

18.8. (b) Here $p(n - 2)180°/n = 180°$, or $p = 1 + 2/(n - 2)$.

18.9. See, *e.g.*, H. Steinhaus, *Mathematical Snapshots* (New York: Oxford University Press, 1950), and M. Kraitchik, *Mathematical Recreations* (New York: W. W. Norton, 1942).

18.10. (c) The ellipse $x^2/a^2 + y^2/b^2 = 1$ can be obtained from the circle $x^2 + y^2 = a^2$ by a transformation in which the ordinate of the ellipse at each point is b/a times the ordinate of the circle at same point.

19.1. Let P and P' be two planar pieces satisfying Cavalieri's first principle. Let $f(x)$ be the length of the chord of P at distance

x from one of the including parallel lines. Then $kf(x)$ is the length of the corresponding chord in P', where k is some constant of proportionality. Denoting the areas of P and P' by A and A' we then have

$$A = \int_0^h f(x)dx, \qquad A' = \int_0^h kf(x)dx,$$

where h is the distance between the two including planes. It follows that $A' = kA$.

19.2. See Problem E 465, *The American Mathematical Monthly,* Jan. 1942.

19.3. Divide the hoof into two equal parts by a plane p through the axis of the cylinder and let A be the area of the resulting cross-section of the hoof. Construct a right prism having as its base a square of area A, the base lying in the plane p, and having an altitude equal to the radius r of the cylinder. Cut from this prism a pyramid whose base is the base of the prism not lying in p and whose vertex is a point in the other base of the prism. This gouged-out prism may serve as a comparison solid for one of the halves of the hoof.

Or one may choose for a comparison solid for half the hoof a solid whose base is a right triangle of legs r and h, whose altitude is r, and whose upper base is a line parallel and equal to the hypotenuse of the lower base.

The volume of the hoof is $2hr^2/3$.

19.5. Let AB and CD be two line segments in space such that (1) $AB = CD = 2r\sqrt{\pi}$, (2) AB and CD are each perpendicular to the line joining their midpoints, (3) the segment joining these midpoints has length $2r$, (4) AB is perpendicular to CD. The tetrahedron $ABCD$ may serve as the comparison polyhedron.

19.6. Place the torus on a plane p perpendicular to the axis of the torus. Take for comparison solid a right circular cylinder of radius r and altitude $2\pi c$, and place it lengthwise on the plane p. Cut the torus and the cylinder by a plane parallel to p. The section A in the torus is an annular region of outer and inner radii a and b, say, and the section A' in the cylinder is a rectangle of length $2\pi c$ and width w, say. Now

$$A = \pi a^2 - \pi b^2 = \pi(a^2 - b^2) = \pi(a+b)(a-b) = 2\pi c(a-b)$$

and

$$A' = 2\pi cw = 2\pi c(a-b).$$

Since $A = A'$, it follows that the volume of the torus is equal to the volume of the cylinder. That is

$$V = \pi r^2 (2\pi c) = 2\pi^2 r^2 c.$$

19.7. (a) See Figure 82.

19.8. (b) Any section, being a quadratic function of the distance from one base, is equal to the algebraic sum of a constant section area of a prism, a section area (proportional to the distance from the base) of a wedge, and a section area (proportional to the square of the distance from the base) of a pyramid. Thus the general prismoid is equal to the algebraic sum of the volumes of a prism, a wedge, and a pyramid. Now apply part (a).

19.8. (c) Let $A(x) = ax^2 + bx + c$. Show that

$$V = \int_0^h A(x)dx = h[A(0) + 4A(h/2) + A(h)]/6.$$

19.9. (b) The section of the ellipsoid

$$x^2/a^2 + y^2/b^2 + z^2/c^2 = 1$$

formed by the plane at distance z from the xy-plane is the ellipse

$$x^2/a^2 + y^2/b^2 = 1 - z^2/c^2$$

having semiaxes

$$(a/c)\sqrt{c^2 - z^2} \quad \text{and} \quad (b/c)\sqrt{c^2 - z^2}.$$

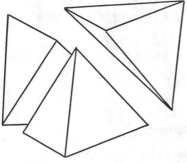
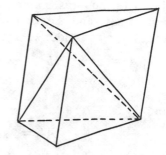

Fig. 82

The area of this ellipse is

$$\pi ab(c^2 - z^2)/c^2,$$

showing that the ellipsoid is a generalized prismoid. We find $V = 4\pi abc/3$.

19.9. (c) $V = 16r^3/3$.

19.10. No. Consider two square prisms P and P' of the same altitude, where P is a right prism and P' is an oblique prism having two of its faces perpendicular to the base.

This exercise warns us that intuition may sometimes lead us astray. One might think of the lateral areas of P and P' as each made up of a number of square loops of thin string (forming perimeters of sections of P and P'). Since the loops are all of the same length and there are the same number of loops about P as about P', one might conclude that the lateral area of P is equal to that of P'.

20.1. (a) Orthogonally project the ellipse into a circle. Triangles of maximum area in the ellipse correspond to triangles of maximum area in the circle. But a necessary and sufficient condition for a triangle inscribed in a circle to have maximum area is that the triangle be equilateral, and a necessary and sufficient condition for the triangle to be equilateral is that its centroid coincide with the center of the circle.

20.1. (b) Orthogonally project the ellipse into a circle. The variable chord of the ellipse corresponds to a variable chord of the circle which cuts off a circular segment of constant area. But the envelope of this variable chord is a concentric circle.

20.3. (b) See, e.g., Nathan Altshiller-Court, *Modern Pure Solid Geometry* (New York: Macmillan, 1935).

20.4. Use the midpoint and distance formulas.

20.5. Place the right triangle with its legs along the positive axes of a rectangular Cartesian coordinate system.

20.6. One might place the triangle with its base along the x-axis and its opposite vertex on the positive y-axis of a rectangular Cartesian coordinate system.

20.7. (b) Yes.

20.8. Take the x-axis of a rectangular Cartesian coordinate system along the line joining the two rocks, with the origin midway between the two rocks.

20.9. Represent A, B, C by $(0, 0)$, $(3, 0)$, $(0, 3)$ of an oblique Cartesian coordinate system.

20.10. Take the locus of A as the y-axis and the locus of B as the x-axis of a rectangular Cartesian coordinate system. If $AP = a$ and $PB = b$, the locus of P is the ellipse $x^2/a^2 + y^2/b^2 = 1$. (This yields the so-called *trammel* construction of an ellipse of given semiaxes. Machines for draftsmen are available that construct ellipses in this fashion.)

INDEX

Abel, N. H., 179
Abû Kâmil, 162
Abû'l-Wefâ, 104
Abundant numbers, 111
Adelard of Bath, 72
Aeschylus, 90
Ahmes, 163
Alexander the Great, 70
Algebra, etymology, 160, 161
  rhetorical, 126
  symbolic, 126, 130, 131
  syncopated, 126, 128*ff*, 130
Algorithm, etymology, 160
Aliquot part, 110*n*
Al-Khowârizmî, 160, 162
  Hisâb al-jabr w'al-muqâbulah,
  160
Almagest (Ptolemy), 75, 99
American Indians, 1
Amicable numbers, 110, 111, 121
Analytic geometry, 218*ff*
Anchor ring, 213
Antaeus, 194
Antiprism, 199
Apollonius, 219
  Conic Sections, 75
Apple, 199
Archimedes, 9, 20, 74, 75, 83*ff*,
  177, 207
  area of sphere, 85
  burning-glasses, 84
  compound pulley, 84
  death, 85
  defence of Syracuse, 83, 84
  first law of hydrostatics, 85

Method, 20, 85, 86, 87, 92
  method of equilibrium, 86*ff*, 92
  On Floating Bodies, 90
  On the Sphere and Cylinder, 85,
    86, 88, 90, 91
  problem of the crown, 84, 91
  Quadrature of the Parabola, 92
  quadrature problem, 93
  spiral of, 93
  theorem of the broken chord, 94
  tomb, 88, 89
  treatises, 85*ff*
  trisection problem, 92, 93
Archytas, 54
Area, ellipse, 208
  parabolic segment, 92
  sphere, 85
  spherical zone, 91
Aristarchus of Samos, 96, 99*n*
  *On Sizes and Distances of the
    Sun and Moon*, 96
Aristotle, 45, 195, 196
  wheel of, 202
Arithmetic, 110
  etymology, 128
Arithmetic in Nine Sections, 10
Arithmetic mean, 12, 14
Arithmetica (Diophantus), 75,
  117*ff*, 128
  problems from, 118, 119
Arithmetica logarithmica (Briggs),
  184
Ars magna (Cardano), 173, 175,
  176
Artificial numbers, 185

INDEX

ccre

Grating method, 192
Greatest Egyptian pyramid, 13
Greek Anthology (see Palentine Anthology)
Greek mystery, 17
Greek numeral system, 129
Gregory, D., 132
Gresham, Sir T., 187
Gresham College, 185, 187
Guldin, P., 211
Gunter, E., 185, 188
  chain, 185
  cosine and cotangent, 185
  logarithmic scale, 188

Hadamard, J., 116
Hambidge, J., 49
Hamilton, Sir W. R., 90
Hardy, G. H., 46, 90
Harmony of the Worlds (Kepler), 198
Harriot, T., 131
  Artis analyticae praxis, 131
Heath, T. L., 128
Heiberg, J. L., 86
Heinzelin, J. de, 2
Heliocentric theory, 96, 196, 197
Hercules, 194
Heron (Metrica), 75
Heronian mean, 12, 14
Hieron, King, 83, 84
Hieronymus, 22
Hilbert, D., 70n
Hindu-Arabic numeral system, 160, 168
Hipparchus, 97
  table of chords, 97, 98
Hippasus of Metapontum, 53
Hippocrates of Chios, 71
Hisâb al-jabr w'al-muqâbalah (Al-Khowârizmî), 160
Hobbes, T., 73
Hoggatt, V. E., Jr., 166
Holzmann, W., 117
Hoof, 212
Hyperbolas of Fermat, 221

IBM 7090, 112
Ideals, 120
Illustrated London News, 155
Incommensurable magnitudes, 45-46
Indeterminate analysis, 166
Indeterminate linear equations, 124
Inductive geometry, 9
Insertion principle, 92, 93
Integral calculus, 83, 85, 88, 195, 198, 212
Irrational numbers, 44
  decimal expansion of, 49
  $\sqrt{2}$, 44-46
Ishango bone, 2
Isogoge ad locus planos et solidos (Fermat), 221

Jefferson, T., 34
Jesuats, 206
Jesuits, 206
John of Palermo, 167
Jones, W., 132
Journal für Mathematik, 5

Kamayura tribe, 3
Kepler, J., 83, 186, 194, 197ff, 207, 208
  Harmony of the Worlds, 198
  laws of planetary motion, 194, 197, 198, 203
  polyhedra, 199
  principle of continuity, 199
  Stereometria doliorum vinorum, 198
Kepler-Poinsot star-polyhedra, 199
Keyser, C. J., 76
Khayyam, Omar, 151, 172
  geometric solution of cubic equations, 152-154
  Rubaiyat, 152, 155
  tomb, 155
Khwajah Nizami, 155
Knox, J., 186
Kramp, C., 132
k-tuply perfect numbers, 112